I0337124

"People, without religion, we'd be looking "UP" for spacecaft when contact happens. Primitive man said heaven is "UP" and they created us! Science compels us to look for this evidence. Why? Because it matches today's knowledge. We live "UP" and create too. Heaven is Space, UP to us as well."
—*Mike Brumfield*

HEAVEN IS SPACE...UP!

MICHAEL

Mike Brumfield

"If the evidence changes, so must the theory."
—*Grissom, CSI of CBS*

"The evidence hasn't changed! The 'story' did."
—*Michael*

Heaven is Space . . . UP!
By Mike Brumfield

Copyright © 2007 by Mike Brumfield

All rights reserved. No part of this book may be reproduced or transmitted in any form or by any means, electronic or mechanical, including photocopying, recording, or by any information storage and retrieval system, without the written permission of the Publisher, except where permitted by law.

ISBN 978-0-9740390-9-1
Library of Congress Catalog Card Number (Applied For)

Retail $13.75 plus $2.50 shipping.
Expect delivery in two to three weeks.
To order call 931/657-5815.

First Edition, February 2007

Did flying saucers create religion?

Dedication

I dedicate this book to Barbara Walters and her special, "Where is Heaven?"

The following is my scientific answer, based solely on evidence, to Barbara Walters and her famous question/interview that was recently aired in August 2006. The question that she asked of all the world's major religious leaders: "WHERE IS Heaven?" Of course she said we all know it's up from the onset of every interview. But do we really? Well, her esteemed panel didn't. None of them took this seriously, and the majority of them said it was a spirit or dimensional world. They certainly didn't think it was up in the sky and never referred to the sky as space. The funny thing is, that one of them even said it wasn't real. He said it was a metaphor. You know like the good and evil in "US" all.

Heaven Is Space... UP!

What! Who does this guy think he is? How dare he judge us and say we all have evil in us. That's not for him to say and remains to be "SEEN"! Maybe he needs to confess? But what else can we expect from a RELIGIOUS scholar?

Religion is synonymous with judging/RIGHTEOUSNESS/hypocrisy (well maybe not all)! Well, what can we expect? For a start, some day, is hopefully science. Science would let the evidence be the judge and not him. It would have no feeling with its verdict, either. If scientific evidence can answer this question, will Barbara even see it? This remains to be "SEEN" as well.

I'm begging her to do a follow-up show with scientists who acknowledge the ancient flying saucer and alien artwork evidence. Maybe the answers will surprise her and agree with the universal ancient "UP" evidence of heaven in the Bible. Barbara laughed after each interview that none of them said "UP," like ALL ancient writings say. And I did, too. The evidence clearly says it is up, and she even mentioned that we all still look up today. I'm sure her laughter wasn't because she saw the irony of this scientific reality—and she tells us so at the end of the show—of Heaven being "UP" and us going "UP," like I now do. I'm sure, because she never mentioned it and at the end of the show even said herself that she still wonders if we will ever find Heaven. Well, I would love to show her as a scientist presenting evidence of why it is up, that we have already found it, like I wrote on my fourth book's cover, *SCIENCE FINDS Heaven!* But will she let science do what religion couldn't or is she religious and "CAN'T SEE THIS SCIEN-TIFC MATCHING EVIDENCE"? Will any religious person see the shocking matching evidence of SCIENCE AND RELIGION'S "UP" FACT OF LIFE!? Take away the magic spirit factor, and I think they could. I think some will, like I finally did. I did it by proving it to myself. I've seen no proof of instant creation, magic spirits or anything living that can't be killed! PROOF IS SCIENCE and "Heaven" began irrefutably "UP!" If you don't "BELIEVE" that Heaven is "up," then please give my answer and evidence a chance. Please, Barbara, for our children's sake. Religion has us on the brink of global nuclear war!

My answer is scientific hard provable evidence.

Heaven Is Space... UP!

It is the reason I titled this book *Heaven is Space ... UP*—and after all Barbara does agree!

At least in premise. Please laugh here, everybody, and for me, take life easy today. Let religion go. Sing along: "Let it be." Let's all sing! Singing is good for the soul, and I'm a soul man! Oh. how I love to sing. Let's all sing, please? It will bring peace to our valley. I can promise you that. Oh, by the way, don't ever "promise a rose garden." Somebody's life ain't so rosy today even if yours is! And please don't ever forget that. Now, let's all have a coke, Lionel Richie, and teach the whole world to sing in perfect harmony! Please? Let it be, let it be!

Author's family struggle now turned world-wide!

(Please bear with my sometimes rambling passionate brainstorming)

Did Flying Saucers Create Religion?

SPACE IS "UP" AND RELIGION IS A STORY OF PEOPLE WHO LIVE "UP" IN THE SKY! WE LIVE "UP" IN THE SKY!

I was shocked at my lifelong primitive spirit thinking when this scientific matching reality finally dawned on me. Moreover, these two things agree, living in the sky and creating. We're doing both! I truly understand how my life changed from a boy whose religion teaches/believes the universal spirit story, to being a scientist. My parents' religion started with the commandment to "have no other gods before me" yet says he's the only one. And then, to add insult to injury, we sanctified our

Heaven Is Space... UP!

prayers to "HIM" with an Egyptian God named Amun/"Amen." What? No Way! What a contradiction and a joke! I understand why I didn't see this scientific reality as I so well know now. My parents weren't teaching me science. I finally know we must conquer space to save ourselves from each other, let alone natural disasters. But do they? No!

This is as much a scientific fact for us today as it was for the ancients who gave us religion thousands of years ago. But how can this be? How could they know this then and we not know it today? They weren't scientific and certainly were brutally primitive by today's standards. But even more importantly, they were religious followers of "heavenly creators!" WHAT! Who the heck are these people? This can't be a coincidence. We need to go "UP" and find them—they say we came from "UP", and we are going "UP"!

THIS IS INDISPUTABLE MATCHING SCIENTIFIC EVIDENCE in solving our mystery! There must have been contact with life that had conquered space. Could these religious heavenly occupants be real flesh and blood beings who have always existed like space itself? Space Dwellers? If so why the hell would they stay away?

And we surely want to know if they look just like us. I mean, after all, it is a given that no one has been able to take us to their leader or make them show up. This I find applies to both religion, scientist, ufologist or atheists who think life does exist elsewhere in the universe. Huh, think about this. Are all these really coincidences? Are we so naive to think we're the only ones in this INFINITE universe? Maybe we are an unstoppable scientific creation. This certainly supports and explains their absence more than them being spirits. Can't we just imagine the scientific possibility of discovering new evidence to solve these questions? Science does. And what if they don't look like us and are ugly? This goes even further to explain their absence. It supports the lust for power "Fall from Heaven/"UP" story." We all lust for beauty. Could this be

Heaven Is Space... UP!

Hell? The evidence confirms it. HELL IS HAPPENING RIGHT NOW! BABIES ARE BEING RAPED! WE ARE "BEAUTY ADDICTS!"

Today DNA is solving cases that were once thought unsolvable in our past. Science will likely conquer this and much, much, more—possibly even immortality itself, through nanotechnology, cloning, DNA duplication, and even resurrection itself. Maybe even storing memory or even finding that it is infinite anyway within the atomic universe! "WHEREVER THE SUN SHINES LIFE WILL EXIST." *(See ancient Chinese disk!)* Even if this sounds unreachable to you then at least realize that space is our final frontier! And we have reached it. We live in a space station. People, Space is our final frontier. SPACE, THE FINAL FRONTIER! Captain Kirk made this the most famous universal simple scientific fact that we know today! All children are looking to the stars for our future survival. All religious adults look "UP" to the stars. Coincidence? Not if you consider the flying saucer evidence/future reality. And did he sum it up best by saying "to boldly go where no man has gone before"? Is he absolutely sure that no man has gone there before? Are any of us? I'm not. I have déjà vu, and the teaching of reincarnation exists. Besides, religion teaches prophecy (predicting the future, that we could have done this before). Oh, and don't forget "immortal" spirits too.

But what is a spirit? Could it be a body that can convert its mass to light, or could it be like a hologram? Could it be a form of communication to a primitive species in which it needs to manipulate because of its subservient use and their flesh-and-blood reality? It makes sense, if they're real like us and don't want to be hurt. Could it be something that we just still don't know what the heck it is and that's the reason we have ghost busters/science?

If we don't know then let us investigate further, please? I think that any logical human on Earth would find it hard to believe that we are alone. I think that this commonsense scientific mentality is

Heaven Is Space... UP!

understandable in looking into an infinite sky full of circles just like our own. In fact, being "alone" is not possible in an "infinite" universe. Please let yourself start thinking infinitely, as both science and religion confirm this possibility. It is the foundation of both. The atom and God are both infinite. And so far, the universe is infinite too, until proven otherwise. This is a "Beautiful" fact of life. In fact, it is logical to assume that life is watching us even if it isn't making contact with us.

It is also logical to conclude that space has already been conquered by people who would be light years ahead of us in technology. If the universe is infinite then it is a given that they are! To give them numbers like light years is a contradiction, an oxymoron. Religion says this very thing. They have no time! If the "FINAL" frontier is in fact conquered then reincarnation makes sense if this is our/their incurable disease/lusting for scientific human creation. This is where they would not be able to co-habitate with their "former" heavenly inhabitants.

MICHAEL

Carl Sagan let us know like Einstein that matter/energy can't stop existing. We can't create it, so looking for the beginning of our universe is like buying the beginning of creation in every religion: then you find out "Heaven" doesn't have time! We can't create what makes up atoms!

And the bible says we are Adams! People, come on! This mystery is all about "Why the hell would they stay away?" (Fermi's Paradox). And in my quest to find that answer, I was shocked to find that it may have to do with "what the hell they look like"! Let's stop being afraid of this unknown. They aren't hurting us, and if someone says they are then let them scientifically prove it. "If not, then they're the good guys!" There is ancient and modern evidence to prove they don't look just like us! Frankly they're ugly! And the writings prove that they are real flesh-and-blood people who can be hurt, killed, and have sex with us! What? No way!

Most of all, there are plenty of references to these divine encounters that show the person was

Heaven Is Space... UP!

horrified upon seeing this angel or God! Why am I not afraid of the evidence? I guess because of my experience with this happening to me when I was only six. It's in my second, third and fourth books. But were they gruesomely ugly and what we embellished to now call them monsters today? The evidence must confirm this, and it must be ANCIENT ART for it to be so!

I am not afraid of monsters, nor the dark. I grew "UP." I am afraid of being ugly and dieing. If I get scared in the dark, it is because I can't SEE! I guess I'm like this because the Jehovah's Witnesses, my parent's religion, couldn't answer these questions for me. Worse yet, they scared the hell out of me with demons and devils who are spirits that I can't SEE! Religion terrorized me, taught me nothing and turned me to science. Science gave me a whole new understanding of religion! I wanted answers. As a loving parent, I want to help all children from being tortured by ignorance, religious terrorism, and DEATH.

MICHAEL

Science proves that our atoms always exist! Scientific evidence is our future, and we shouldn't be afraid of challenge! And it, like our children, should be nurtured for the sake of their future! Here's to the future. The scientific religious evidence that you are about to read and see has persuaded me to be scientific about religion's heavenly occupants. This evidence that I have discovered and still am learning, in my quest for answers about our origins, has changed me forever. It made me a scientific believer of these mysterious "PEOPLE "UP" IN SPACE" and answered the torturous question: "Why would they stay away?" This in turn answered for me: "why the universe isn't perfect and why they can't stop us from killing each other?"

Finally, it has answered the ultimate question: "What do they look like and where did they 'come from?'" Again "come from" is a contradiction to an infinite universe and energy/matter. The following evidence "should" tell you why our mystery exists.

Heaven Is Space... UP!

If it doesn't, then maybe it flies in the face of your religious prejudice as it did with some of my family.

Please don't let words hurt you. We teach children "sticks and stones may break my bones but words will never hurt me." Come on. This evidence is also about helping the brutality of humanity as science represents today. It will speak for itself, without any prejudice on my part. I follow the evidence only as any scientist or neutral seeker of answers should! If we all wouldn't die for our children today then we need to re-think even looking for answers. Can religious "BELIEVERS" see the matching scientific evidence between religion and flying saucer evidence? Scientifically, I already know most can't because of the evidence that exists now. They can't give up the "magic"/omnipotent god. Most religious people don't know where Heaven is and believe in spirits, without seeing proof. So seeing flying saucers won't change most of them.

My brother Brian said it best in a recent interview for my new DVD, "Why The Blank Don't They Care?" I was getting his eyewitness testimony of seeing a silver-disk in the sky above a commercial jet while building my billboard (see back cover of this book) last August. I asked him if he thought seeing one will help religious people. He said he asked a preacher this, and even the preacher told him that he would still believe in his God! Even worse, "He" has to come back on the clouds "riding a white horse." No, this can't be happening! Please somebody slap me! I've already asked my family the questions you are about to read in my struggle. If their reaction is any indication, then, indeed, their religious one-third (those cast out of Heaven) will be cast again onto another "lake of fire"/body/Earth to infinitely be religiously doomed. Coincidentally, to a final test of NOT BELIEVING THE SCIENTIFIC EVIDENCE SO OBVIOUSLY ENTRENCHED IN EVERY SOCIETY ON EARTH as a result of non-cohabitation. All this in spite of looking "UP" and giving God a

Heaven Is Space... UP!

halo symbol that looks like a flying saucer.

We are ruled by religious leaders who all look "UP" to Heaven, but believe in spirits. This is chiseled in stone. I've asked them, my family and many people at my book signings or lectures, why? I'm not surprised to find that none answer scientifically or don't remember they were taught this ignorance and obvious contradiction. Why would spirits come from "UP?" Let alone be riding white horses! There parents don't know why, either! Only the scientific will see this matching "UP" evidence. Scientists will let the evidence and lack of it say who these gods are. What's really scary is that our world leaders are religious. They believe in Heaven, kill children just like their god, and have NUCLEAR WEAPON capabilities to destroy the world! That's the scariest thing of all. How could this be any scarier than the Biblical Hell? People, this must be Hell!

When I press the issue that "Jesus"/Yeshua himself saw religious people didn't know where Heaven

MICHAEL

was and would kill too, they would get very angry. I have also been attacked and received religious death threats. But, I still want to ask the world: WHERE IS HEAVEN AND WHY DON'T YOU KNOW IT'S SPACE? SPACE IS "UP"!

Yes, I would especially love to ask Captain Kirk this very thought-provoking question. If nothing else, to see how much his wonderfully futuristic-based scientific phenomenal show influenced his scientific awareness of "OUR" religious world around him. Did he and does he see the matching scientific evidence of flying saucers to heaven's fiery chariots? They are both "UP" in the sky and this is where they came from. And of course, this is where we are going . . . "UP." We all know that conquering space is the logical course of events for our species to survive. Or at least I thought so, until I scientifically examined religion and its adult followers. Most don't think scientifically. They don't know this fact and, yes, I'm shocked that they believe God's magic will supersede science. Only God will save

Heaven Is Space... UP!

us! And of course, they all want you to send God money because he owns all the wealth and is our ultimate lien-holder. To prove you believe this to them you should give to God. But of course, again, through "Pat" who handles everything.

I just heard Pat Robertson say this on the *700 Club* at 10:00 a.m. this morning. To my horror he was on NBC. No! This is influencing our kids, people! How can this be? He's the "pResident's" friend. He is also the same guy who preaches revenge instead of forgiveness, totally the opposite of whom he represents. Just like the President! I can't stand how both use science when they want and use religion when they don't. How can we let religious people use science when they want and then attack it otherwise? This is the height of hypocrisy. Of course I know how. Religion rules our White House! You will see why this is so when you view the *J.W. Magazine* using science to prove God exists through intelligent design, but then saying science is not to be trusted either. Crazy, huh? Worse yet, they say something can't come from nothing! What they don't see is that's exactly what their story is. Their "invisible"/spirit God just thinks it into instant existence! Although, when it comes to all children, religion doesn't rule here. They want visible, scientific proof. They know, because of science, that we need to get off this planet and conquer space.

But, what if there was proof that it already is, contrary to Captain Kirk's popular sound-bite's "to boldly go where no man has gone before" made oh-so-famous by *Star Trek*? And if there was, why wouldn't he or any of his cast members have "seen it" by now, especially the logical one, MR. SPOCK? Could they also believe in the magic invisible God who lives in the sky? Would evidence change them? You should already know the answer. It's already written in the religious "EQUATION", two-thirds will and one-third won't. If they did "see it" we'd already know—they're famous. So, just imagine that it's you who discovers flying saucer evidence and

Heaven Is Space... UP!

sees this similarity between "it" and religion. And how hard it would be to get respect. Well, this is my story. Now, just imagine trying to get an Intelligent Design religious proponent who says "SOMETHING CAN'T COME FROM NOTHING." That is exactly what his God story is—to REALIZE, that's what it does and is. His God is invisible and just speaks things into existence. VOILA: SOMETHING CAME FROM NOTHING! COME ON, RELIGIOUS PEOPLE, GET SOME SCIENCE/COMMONSENSE. If you don't know where the hell God is, what he is, and why the hell he would stay away, then just say I don't know and start researching it scientifically. If you were buying land you would, wouldn't you?

I mean, after all, nobody buys swampland in Florida without going to see it! Isn't your God worth proving? Well, what if we could present scientific evidence to prove this very thing or not and that space is inhabited? Before you think this is impossible, let's look at our history. We all know the

MICHAEL

old saying: "To learn about our future is to look at our past." That's the only way to see where we are going, to know where we came from. The historical evidence might surprise you. Our history began with religion universally saying that Heaven is "UP" and inhabited. Science agrees with these two basic simple premises. We send space-craft "upward" to the "HEAVENS and monitor them every day in hopes of living there, just like religious people look "UP" to the "HEAVENS" every day in hopes of living there as well. AGAIN MATCHING EVIDENCE! (Read about J.W. convincing most followers and my mom to stay on earth in paradise; they are afraid of Heaven because they don't know what to expect. They've always known this. Talk about fear of the unknown!) But, why hasn't Kirk or his co-stars come out and disclosed any awareness of this "coincidental" matching evidence. Recently, we've even spread Scotty's ashes into space. He was famous for beaming people aboard their craft. This would scientifically explain religion's "magic"

Heaven Is Space... UP!

teleportation through science if it was possible. We have successfully done this with gas atoms and the future looks "bright" for carbon atoms as well!

We are also levitating things, which is another universal religious teaching that is being done scientifically. However obvious this matching scientific evidence, does it change the minds of the Japanese people who lead the way in this technology? Shouldn't they or the Hindus be the most likely to see this "coincidence"? Does it change their president or our own? NO, it hasn't!

So, it's easy to understand how Kirk and his co-stars are still enslaved by primitive man's magical spirit God teachings, if they are. It's painfully "CLEAR" how this scientific blindness exists when it comes to religion. Could this matching scientific "UP" evidence alone prove that space is, in fact, inhabited? I think it is possible only through the scientific examination of religion's "spirit" history. We all know how stories can get exaggerated from the same exercise of story-telling in school. All religions say Heaven is "UP"! How could this be, if the ancients were primitive and these beings are spirits? Spirits wouldn't need direction. And for that matter, neither would today's quantum dimensional theorists—who like religion say that dimensions are a fact—need direction either.

DIMENSIONS AND SPIRITS ARE NOT PROVEN FACTS! If they don't admit to this then forget it. This is not debatable; only one scenario "MATCHES" the ancient evidence. "RELIGION" is a "UNIVERSAL STORY OF PEOPLE LIVING IN THE SKY"!

WE NOW LIVE IN THE SKY on the International Space Station. This matches! If you think this is just a coincidence, that primitive people would have this scientific knowledge, then please look at the pictures on the front and back cover again. These are all ancient religious artworks by primitive man who drew what he saw! We know this because he also drew the animal and plant life too!

Heaven Is Space... UP!

The following evidence will disprove the "SYMBOLIC" argument in religion and science, which is just their theory/opinion. It isn't a law. It takes "MATCHING SCIENTIFIC EVIDENCE" to establish laws of science, like the law of gravity! If you throw an apple up it will come back down. These RELIGIOUS artworks are all looking, pointing, and going "UP" into "SPACE." For this reason and one other that you will read about in the last chapter of my story, I titled this last book *Heaven is Space . . . "UP"*. I did it because we say Heaven is "UP" every day; we see "PROOF" of it every day on television around the world; we read it in all the religious text; and we even look "UP", but we don't really believe it's "UP" there "in" space. My religious family struggle will epitomize Heaven's ultimate confusion later, for those of you who don't get it that we're in Heaven/Space/UP now and that Fermi's Paradox is the real question to be answered.

Well, I mean, besides the obvious ultimate reality that we will "FACE" upon their return. Which is: "What will their 'FACE' look like?" Remember, though, I think you will "CLEARLY SEE" why one causes the other to exist, with we humans who love outward beauty! And if you don't know "FERMI'S PARADOX" term then science and education is the only thing that will really "TEACH" you—not religion. It is the same religious question that they can't answer with proof as science does.

Anyway, if this "POWERFUL UP" evidence was the only matching scientific coincidence I had to answer our mysterious loneliness with, then maybe it wouldn't be possible to accept it as a fact. But there are more scientific examples of matching evidence. Plenty more! Like the "GOLD" halo symbol, given to these heavenly people as a result of them not being with us and again coincidently "UP" above their head, resembling a flying saucer. (Cave art on back of aborigine god in flying saucer going "UP" protected by gold halo!) YES, A FLYING SAUCER, and yes, there are plenty of ancient

Heaven Is Space... UP!

artworks showing more all over the Earth! Besides, I filmed them and found my footage to match these ancient artworks as well as my buddies in phoenix. And it is on Dan Aykroyd's video right now called *Dan Aykroyd Unplugged On UFOs*. (Check out our new video, *Why the Blank Don't They Care*.) Yes, he and many others—including Jimmy Carter, a former president, and also another president, Ronald Reagan—saw one. I don't have to ask how they can't entertain the idea that maybe these craft started religion? Because I already know. They all believe in the primitive, magic spirit/God story!

Even more important, though, than all these famous people seeing them, too, the footage and still image MATCHES the so-called HATS of Easter Island. And they are looking "UP!" It is shown on the front cover. Come on, people, this is good scientific proof that religion was started by bald-headed people in flying saucers "UP IN THE SKY." This also matches other cultures' "hats" and baldness, which is well-established in religion, like

China's rice hats and Buddhist monks. So is uniformity! IT ALSO MATCHES THE UNIVERSAL TEACHING THAT Heaven IS "UP" and they're all equal!

The Easter Island statues are all bald and look the same, like Buddhist monks and the Olmec gods. They all look "UP" to Heaven. For "GOD'S" sake, the Owl Man on the back cover is pointing "UP" to the sky as well and has a huge bald head and big eyes! The aborigine God below it shows the bald alien in a saucer going "UP"! Come on, people, again! This is hard scientific evidence of what primitive man's God looked like and where he is! If you are religious and this doesn't sway you then nothing will! You will be forever allegiant to your religious faith, which is the absence of evidence and your saying that you don't need it.

Well, I'm not saying you need it either! But don't try and convert a scientist to your magic spirit/invisible God when you have no proof of him, don't know where he is and certainly—it goes with-

Heaven Is Space...UP!

out saying—don't know what he looks like, because he's invisible! This will make you look insane. Funny, huh: no "COMMON" sense! Religious killers get off on insanity pleas like Andrea Yates, but we don't see the insanity in the religious God killing his children. Now that's crazy as hell, and religion created both the killer and the insanity defense for it— and we don't see RELIGION as insane! Even worse, we now see why Andrea got off but the mother who killed because the devil told her to do it didn't. She got the death penalty while Andrea's God defense spared her! What a twisted scientific conundrum, huh? How do we cure it? We can't. I recently told Jehovah's Witness to do one thing for me and that is to read, proving to themselves that their God is a murderer. He kills innocent children in Sodom and Gomorra. They don't even "mind!" If you don't get it by now you won't unless you continue to look at the evidence. THERE IS LOTS MORE!

I like the Olmecs best. The Olmec gods look

the most like the Ros-well gray alien—JUST LIKE HIM AS A MATTER OF FACT—and the statues from Israel found along the banks of the Jordan River! And, yes again, the worldwide Roswell saucer crash exists. But it is nothing like the Washington, D.C., evidence in 1952 of a living president knowing they exist and not making a worldwide effort to find them! This blows me away. And for this reason I made my latest video, which is a follow-up to the "proof" video internationally nominated in the Film Fest for the study of UFOs. It came in second place at the 2006 convention! See, I don't mind not winning. Uh oh, Karma just ran over my dogma, and humility is lying all over the floor.

The follow-up video is titled *Why the Blank Don't They Care?"* and has the footage you see on the front cover of my book. I finally filmed them March 6, 7, and 8, 2006, after making the proof video with Jeff Willes in 2005. You can see the proof for yourself. Fourteen flying saucers are clearly visible in the live footage and pictures over the White

Heaven Is Space... UP!

House, and we couldn't stop them or catch up to them with our fastest military jets!

But enough of this POWERFUL SEEING IS BELIEVING EVIDENCE, because, believe me, seeing a picture, video, or newspaper won't change a skeptic. It seems nothing will change a religious follower but themselves. I guess it takes seeing one for yourself—and this ain't easy. You have to look a lot! I finally saw one for my brothers' recently and asked them if they landed and said they started our religion would they believe that. They both said if the evidence was there, they would. Cool. It is! *(See covers)*

For now, let's get back to the gold thing. I always thought that if I filled in our history with all the evidence, revealing that real flesh-and-blood aliens in flying saucers existed, then that is the way I thought it could work. I am also proving that they wanted our gold and created our religious stories. Much like convicting a criminal with DNA, fingerprints, blood, footprints, eyewitness testimony and

MICHAEL

every other piece of matching evidence would. But then again, I remembered O.J. We had all that evidence and he was found innocent. How? How did he explain that his blood was found at the crime scene? Oh, I remember, a racist cop. But is it just a coincidence that his finger was cut, too? Oh, how I wish lie detectors were perfect and everywhere. More importantly, I wish there was a God who could make this world perfect and it didn't happen. But there isn't and that's why I challenge the perfect magic/God story. Most importantly, I hope I have the strength and love to forgive him if it did.

If he did it, then it was a result of "SICK LOVE" like the God story. Really, I deeply feel for Nicole's death and O.J.'s life because the "COLD HARD FACT" is that God kills his loved ones, too, when they don't love him back—or for even far less. People, this love/worship-demanding, killing God story influences everyone. It is the same sickness O.J. could have if he had done it—killing someone because they wouldn't love him! For Nicole's sake,

Heaven Is Space...UP!

the Yates' children, and the perpetrators, too, let's solve this mystery.

I thought my last book about gold's historical connection to God and space exploration would. But it didn't. Anyway, what I discovered about gold is unbelievable! Our history was founded in gold mining. Why? It was clearly used by the ancients for their sacred offering to the "SKY-PEOPLE"/God, and not because it was rare, either. It is abundant. I discovered this scientific fact during the writing of my second book and it also astounded me. I always thought it was rare. What I discovered awakened my scientific mind.

Gold is essential for space exploration. Our astronauts have a gold-covered face visor. The aborigine God on the back cover is protected by the same gold halo over his head—and he looks like an alien! Heck, all the pictures on the cover do! Bald, big eyes, and indicating "UP"! The religious quotation "Heaven's streets are paved in gold" again MATCHES TODAY'S SCIENTIFIC EV-

IDENCE OF SPACE EXPLORATION ONLY MADE POSSIBLE WITH GOLD! WE PAVE/EXPLORE IT WITH GOLD! Come on, people, this was hard scientific evidence that could solve religion's mystery: WHERE ARE THEY? WHAT DO THEY LOOK LIKE? and WHY DO THEY STAY AWAY? I thought this would make my second book not only a world-wide best seller, but also finally convince my family that "Heaven is in fact UP or space" as we know it is today. But it didn't. Neither did all the other amazing scientific matching evidence either.

But, here's a few more matching examples just to prove my point even more. Not that it will change anything, but I'm an optimist and really do hope for the best always. Primitive man was drawing what he saw even if it was not his knowledge and obviously God's like the flying saucers, atoms, and DNA. It is clear that they all play into primitive man's story of his "SCIENTIFIC CREATION AND THEIR SPACE-FARING FLESH

Heaven Is Space... UP!

AND BLOOD REALITY." First, the double-helix coil of the A.M.A. symbol. It looks like DNA and does match the double-helix strand we all know so well today as "the invisible scientific creator of life." If he was scientifically created then he could have seen this symbol at some point, given they used it instead of magic/instant creation. The matching evidence indicates this reality as well as universal snake worship and the same squiggly-lined-looking design all over the Earth. The second is the obelisk, which looks like a rocket to me. And in fact, the known Egyptian origin confirms it to be a weapon of the sun God. A weapon that eerily looks like our nuclear missiles today! Third, the atom looks just like the Jewish star and matches it point for point. I put it on every book.

Our universal religious story is one of pre-destination—a creation known to end in our own demise from threatening the destruction of the earth through war! Come on, people. North Korea has nuclear weapons and Iran wants them. We have just now gotten the capabilities through nuclear weapons to destroy the Earth. Iran wants to wipe Israel off the map! Please help me, people. Do any of us really want to see our children put through nuclear hell? Isn't this hell bad enough?

Listen to me pleading as if it can be stopped. If destiny can be stopped, then it has to start with religion itself. Religion agrees! Don't you see the cruelty of Iran's religious president or our own? Do you really still buy the magic spirit God story with all this scientific evidence of our species being a scientific creation like plastic. We both have an instantaneous occurrence on history's scene. I found that most of my family did believe in the same God as the Iranian president and ours still does too.

They just don't get it. Most people even get angry because of their lack of evidence and just tell me that I'm going to hell for questioning God. I tell them it's their God, not mine. They're the ones with faith, who don't need evidence, and worse yet, don't "mind" God-killing children! No. Please

Heaven Is Space... UP!

don't tell me that this is the way our world truly is. Somebody, please wake me "UP" and tell me that I'm just having a bad dream. Dream, hell. This is a nightmare. Our Earth is literally surrounded in darkness! No, not the outer darkness they call hell! Could it be? No, please?

If it is, then the fact remains that their lack of evidence and looking for it, feeds this monster mentality called religion and its killing God. It creates presidents who kill, too. I can't believe this is real, but it is. Well, I finally accepted the reality that nothing scientific will change a religious person like my mother who has been a Jehovah's Witness for fifty years. She says they don't preach hell, though, just the killing-God thing! I told her that it doesn't make them any better, because they're not. They're all the same! You can't get worse than killing. The sad irony is that she still doesn't know where Heaven is, only that it is a spirit world. And remember she's afraid of spirits. This is crazy! This really frustrated me as the "UP" evidence was so overwhelm-

MICHAEL

ing, besides religion's lack of magic that they claim God has. And it really kills me when they say God's power isn't magic. I don't know what the hell they think it is then. He does just speak things into existence. That's pretty damn magical to me. Nothing sways their "belief." I even gave them example after example of situations that would cause this same magic spirit story if we were to land on a planet inhabited by primitive mankind. He would say we came from "UP". And if we weren't able to live with his uncontrollable brutality, we would leave. This isn't realistic if we couldn't be hurt like a spirit. If we scientifically altered his intelligence to serve as a slave for us, his story would go as follows. His God came from "UP" in a fiery chariot or cloud, is magical, and they were created to serve him. The reason he's alone, is because he rebelled with his "new-found" intelligence. They also would have a slow evolutionary growth until this moment and then civilization would just have suddenly appeared culminating in a very short history of rapid growth.

Heaven Is Space... UP!

Wow, Wow, Wow. SETI, eat your heart out (just kidding, karma)! THIS IS OUR MYSTERIOUS PAST! Even more, their need for gold would also be obvious as they live in spacecraft. Our history started with gold mining everywhere, and it still backs all currency today!

The scary part is that all religions say that we are doomed to destroy ourselves, and they knew it before they gave us the ability to do this and yet did it anyway. But the good news is they also stop this from happening with a world-wide famous "UNIVERSAL" Second Coming ending! And where do they come from? The world's largest religion knows this all too well.

"JESUS WILL BE COMING ON THE CLOUDS FOR EVERY EYE TO SEE!" My mother says this is not only symbolic, but the Jehovah's Witnesses said it already happened in 1914 "INVISIBLY"! Come on! Please tell me that there is one Jehovah's Witness who sees the ridiculousness and danger of this religious fraud. He's invisible with no way to prove it! What! No way! Why aren't they teaching Religion 101? When did the God story start, where is he, she, or it? And most importantly, why the hell would he stay away and torture us with this mystery of mankind's brutality, anyway? WHY, WHY, WHY? If we are afraid to accept the possibility that he, she or it isn't a magic "ALMIGHTY" spirit person then I finally know how this happened and why it will be this way "FOREVER"! Religion breeds a traditional incurable resistance to any scientific evidence that refutes it.

HEAVEN IS "UP" IN SPACE, AND HELL IS DOWN ON EARTH! Space is "OUR" scientific final frontier! This is scientifically logical. Earths make great universal prisons. "Our" is the key word here, and that is speaking as a scientist seeking scientific evidence—evidence that should speak for itself! I've even discovered that the Mayans painted their gods blue because they needed to distinguish them from mortal humans. Blue because they live

Heaven Is Space... UP!

in the "SKY"! It's obvious how all man makes God his race. Mom's God not only lives in the sky, but is a "man" spirit and she doesn't know why "HE" would be. She doesn't know because she gets my point. Spirits wouldn't need to live in the sky, only us flesh and blood humans and anybody else trying to escape the hell of their ignorant, beast-like population would. Just like the aliens with us! Get it!

And they would be in spacecraft as the scriptures say: "Flesh and blood can't enter the kingdom of Heaven." SPACE WOULD KILL US! But does she really "GET IT" or is she just pacifying me to stop this torture? I was torturing everyone, even my buddy who videos flying saucer and yet believes, just like mom, in spirits. What the heck is wrong here? Could I get him to finally look for proof of his spirit world? He looked for proof of saucers! I couldn't, even at the writing of this book. What can I do to get them to prove or challenge religion's spirit world. It has "BRAINWASHED THE RELIGIOUS PEOPLE OF THE EARTH." (Thanks

MICHAEL

for the quote, Brian!)

Hell, I wish religion's magic God did exist. If he did though, I would still ask WHY, WHY, WHY? And we all know the questions. But they don't and won't ask why? Why are they afraid or why don't they care about answers backed by evidence? They care about me and getting me to join their church, don't they? Voila! This proves that they are brain-washed by the church! This is what the church wants! People, Space is the final frontier and we can't escape the evidence laid out in the groundwork of our primitive religious forefathers. It still exist today. WE LOOK "UP" TO HEAVEN! Evidence and scientific knowledge will prevail over ancient religious traditions. Even this is a good scientific piece of matching evidence. The bible says that knowledge will increase all over the earth and finally prevail over religious traditional, superstitional ignorance. Is it a coincidence that this has happened? Don't these Bible-thumpers see this? It is provable by the history of progress and religion's

Heaven Is Space... UP!

story of "the inevitable global increasing of knowledge." It is increasing and can't be stopped.

Now I know why the word "FINAL" created religion's reincarnation tradition. Beast will always remain beast. And like beast, I know what the power struggle is in religion's universal story. I see all of our addiction to outward beauty. This is the power that humans possess, our unique difference in looks. Nature all looks the same, from bees to trees. I know all too well! As a matter of fact, my name is at the forefront of this struggle and it again confirms energy/reincarnation and the way our universe works. The most famous religion of the Jews is waiting on Michael as is some of Christianity, like my parents' religion. That's the reason I use it as my pen name. I use myself as an example in my story. I haven't found anyone who seriously entertains the possibility that I could be him. Well, except for the ones like Brian who want me to prove it. But the only way I can prove this possibility is to solve our mystery with scientific evidence. I can't make anyone show either. Funny again, huh? I'm trying to solve a religious mystery with scientific evidence. And we know how the Jehovah's Witnesses view scientific evidence. They use what they want to support a magical God of instantaneous creation, but totally reject evolution, which is universally accepted. Yet, there isn't one example today, nor in our past, that they can give of instant creation!

I beg them to do it, or at least admit the Earth evolved. Anyway, the current evidence PROVES we are forever doomed to be separate from these heavenly beings that are equal. The evidence clearly shows them to be so because of their looks. According to all religions this separation is because of a desire for power to be greater than God! If Earth is a mirror-image of Heaven as it says, then I know they created our species to give them this power. Plastic surgery is the ruler here. Could they have created our "NEW" species for the power to be greater than one another through outward beauty and uniqueness? Are they this advanced? It's clear they

Heaven Is Space... UP!

can transfer memory from one body to another or possibly transfer the memory cells themselves, just like reincarnation implies. It's also clear that they can be invisible. We are on the verge of doing this, too. (Look at newspaper evidence!) Again, see the matching scientific evidence! WE LIKE WATCHING OURSELVES!

People, we are beauty addicts. I am, and I see it every time I go to RED LOBSTER to eat. We like to look at each other! This story is so scientifically provable that it is pathetic. It's so simple, we even tell our children to not judge a book by its cover. Why can't we see all these parallels in our scientific world today? We are on the verge of all this and even more. Heck, we are even creating new species of life today that wouldn't exist, just like us, without science. Please ask yourself: Do I want to be outwardly ugly? I don't! Humans are cruel when it comes to this. That's the reason I put the Olmec gods in the evidence. They are equal in looks, just like the heads of Easter Island. They are ugly (abnormal compared to humans). This is the reason I use my name as a fictional possibility of being the biblical Michael. I am addicted to beauty—worse yet, to my own beauty! I am no better than anyone else. Our sexual ego or desire for power tortures us! It tortures me! I want help. Scientific humility is my only answer, not judgment.

Love of others is scientific humility—forgiving others and myself of falling prey to this most "POWERFUL" feeling. I can easily imagine being an ugly alien addicted to humankind's unique beauty. I also can easily imagine life doing this "somewhere else" in the universe. Right here and right now. Anyway, back to my name and the reason I use it. Michael the archangel wars with the devil (who is an angel) and his angels, according to the story, and they are cast out of Heaven and down to Earth. I found this to be the biblical prison that we can't escape called hell as well as the human body. If we are an alien then it would make sense of all religions saying this body isn't who we are. They all say this.

Heaven Is Space... UP!

Plus, one person is the creator of any new invention. Was this angel just that, the inventor of human creation? Then, at the end of the story, Michael returns from the "SKY" to war with those of mankind who are destroying the Earth. He then cast them to hell again. AGAIN, INFINITELY REPEATING HISTORY! I propose the evidence to be perfectly clear. This story reflects the possibility of an unstoppable creation of mankind, which they called the devil. This is likely the reason YESHUA/"JESUS" called the Jewish preacher "SATANS," plural. This also could be the reason he called the good guys, "only the father in Heaven." He even called himself evil because he was a man when the apostles called him good.

The evidence is clear that these beings all look the same; therefore one is many, monotheism is polytheism. They are a perfect example of nature's creation. They are just like zebras in this sense many but all looking the same. We have plenty of evidence of these angels/gods breeding with their creation in the mother-goddess statues. This would explain why they are called "FATHERS." There is evidence of aliens with a penis. The head of all these mother-goddess statues are bald and big, like the alien fathers. The bodies are all fat, exemplifying the story in Genesis saying they were doing this because of their beauty.

This makes painfully clear again that the power struggle is all about the power we have from our unique beauty. Human mothers with alien fathers. And voila, we all look different. We have a missing link in our fossil record that reflects the lack of a skull between primitive man's small one and our "OVERLY LARGE" one. This is again beautiful evidence that we are science's creation. We have this evidence. (See fossil chart.) Nothing else in nature does this! When it finally comes down to contact, what they look like will not be acceptable to us narcissist beasts. Yes, I love myself outwardly. Sure, there are some things I would change, but plastic surgery is to expensive. I couldn't do it now, any-

Heaven Is Space... UP!

way, unless I was grotesquely "UGLY." I still don't know if I could, though, as long as my kids weren't repulsed at it.

I want to share money with the world to help the ones who are. They "need" to feel the power of beauty as they are tortured by this dreaded "UGLINESS." NO! I can't believe I just said that! Like I said at the end of my third book: "EVERYBODY WANTS TO GROW OLD. HELL, WE EVEN WANT TO LIVE FOREVER, BUT LET'S 'FACE IT' NOBODY WANTS TO 'LOOK' OLD! LOOKS GIVE US POWER OVER ONE ANOTHER!

You know what, folks? When it's all said and done, we aren't going backwards. Children know this cold hard "ugliness" fact all too well and that space is our only hope for survival. They know this as a result of living in a world faced with daily nuclear destruction. Space is the Final Frontier! Unlike them, little did I know the scientific reality of this fact, in humanity's quest for survival, as I was

MICHAEL

growing "UP". This famous quote by Captain Kirk of *Star Trek* did become a staple for me, though, as a child. However, it only "happened" from an occasional glance on the TV as my dad or mom skipped through the channels in search of traditional Fifties & Sixties westerns or family-theme shows.

I was never really taught science by my religious parents. I now know why. They themselves weren't aware of it either, and this was obviously due to the same obvious non-scientific religiously dominated world and upbringing, just like my own. God is everywhere! This is good old tradition at its best. Religion "preaches" against tradition and yet patriotism makes it null and void as a fact. Besides tradition, it was easy to not be scientific in a pretty primitive state back then. And this is only fifty years ago. Just think of where we will be fifty years from now and how today will look in another fifty. It will be history "REPEATING ITSELF"! Hell, it still is a "pretty" primitive planet by any universal standard today—we haven't conquered space,

Heaven Is Space... UP!

moved into it and still don't see the significance of the universal religious "sky" God story as possibly being scientific. I mean, after all, nobody is healing dying babies right now. If they were we would all know it!

We need to conquer that sky; space is our final frontier and it is obviously where these religious inhabitants live! But why don't we see this when we even have evidence of a historically long UFO phenomenon and they, too, are in the sky as the acronym so "clearly" spells out UNIDENTIFIED FLYING OBJECT. "Flying" being the key word. "RELIGION'S UNIVERSAL HALO" looks like a "FLYING SAUCER"! Why don't they see that tradition proves the ancient origin of Heaven is "UP" AND POSSIBLY CREATED BY FLYING SAUCERS? THEY DO MATCH!

This is irrefutable. Heck, we even show all gods flying like the saucers do! Really, I know why religious people like myself didn't see this matching evidence and it becomes excruciatingly apparent in the reading of my story. So much so, you may even empathize with me at the end, that even in the face of contact this mystery will still continue as it does today. It is infinite like the religious hell story. The reason for this will be easy to see after you're finished reading.

I *can* promise you this! I'm sorry it took me five books to finish this journey. This is the last chapter of my harrowing saga. With it, the God mystery is finished, for me, anyway. I'm okay with the evidence. My thinking is now dominated by my brother Brian's "PROOF" factor, or science. They are one and the same as the Jehovah's Witnesses so eloquently show and then look ridiculous by retracting science as a method of proof (again see evidence). I'm a big boy; I've grown "UP". I know Santa doesn't exist. I can do the same with religion's God, not science's god. I can't believe it took me so long to see their similarities. I believe the scientific evidence that "shows" me what God is and why the world isn't perfect, and why magic miracles don't

Heaven Is Space ... UP!

exist. It has answered for me the logical questions on the cover of my book: Who, Where, and Why?

Scientifically, I am not afraid of the repercussions from my parents' religion. Ironically, I am also addressing the religion that rules the Earth, which also consists of five books like mine. Judaism, Christianity, and Islam. Coincidence? Like my parents, I too believe the evidence proves the God story to be a scientific fact. I "BELIEVE" the evidence without the magic. There's no proof of it so far. I have too, based on this. Therefore, I don't have religious questions for their God anymore.

Science has given me the technology and desire to find what I was looking for, religious answers. I am a scientific believer in the God evidence without the instantaneous magic factor and evidence that will scientifically rewrite it. I didn't become a scientist overnight, and in fact science never ends. We will agree with the evidence if it warrants rewriting itself. Competition is non-existent where evidence is concerned. The competitive na-

MICHAEL

ture hurts all fields. Like religion, I knew very little about science, but I certainly knew nothing about religion. Not to mention the fact that you spend the first twelve years of your life just slowly learning about what life is and who we "humans" are, but I really had no clue of anything. I was a kid.

The answer for this is simple. Life is a painfully slow learning process. My parents didn't know "why," and their parents didn't know either. I quickly understood tradition. The questions on the cover of this book are simple questions. My parents were childishly simple with their humble answer, "I don't know the answer, Mike." I could've dealt with that but yet a commandingly complex admonition always followed: "But there are some things you just *can't* understand because 'God' works in mysterious ways." What the hell was that all about? There's no way this can be true! I was completely unsatisfied religiously as well as scientifically. I just wasn't scientifically taught about religion! And when it came to science, well, school was slow like I said and I

Heaven Is Space ... UP!

obviously wasn't "gifted" or it wouldn't have taken you this long to read about my story.

When I did go looking for religious answers elsewhere some twenty years later, I was repeatedly told this same thing by the rest of them and I am still scientifically examining "all the knowledge in the world." And, man, is it complex! It is what this confusion is all about. My life! The world and its origin. I mean, after all, I came from "it" and I am in "it." So what the hell is "it"? If I came from "it" how dare anyone to tell me that I can't trace "it"? As a matter of fact, I am "it."

I love hunting. It was a sacred family tradition. Except this became a hunt unlike any other. I was endangering my life for "it" and it was my life. I wasn't taking a life; I was giving one back. I was essentially finding "me," and according to my mom and dad, "God" made me. I had to know where Heaven was and ironically, laughed at the idea of it being "UP" just like every other religious person. This is so funny, as you will read about in my first book *The Two Witnesses and the Religion Cover-up*. I didn't know where Heaven was and what God looked like. I wanted to talk to him, her, or who the hell it was that could give me some answers. First of all, a NO-BRAINER question: WHY? Why the hell would anybody stay away from us and cause this torturous disease of humanity, our mystery and religion's birth? Why are we alone and why do we have such a short history and why are we so God-awfully, brutally primitive. Wow, now I finally knew where the "God-awful" saying came from. Funny huh, always learning.

But this wasn't fun, these nagging questions, pain and death of children. Hell, these questions alone were so scientifically frustrating for me. At first I didn't even dare, but I did finally ask religion the merciful question, "How could a loving God not stop a baby girl from being sodomized by her savagely brutal and certainly horrible, evil, sex-crazed, mentally deranged supposed "father"? Forget about that—well, I wish I could, but

Heaven Is Space... UP!

I can't. I learned to live with it, like everyone else. Those who couldn't commit suicide. Only we humans commit suicide.

Finding ancient alien evidence as the title of my second book says, *Aliens Gold Tenth Planet*, I asked the question that was so beautifully illustrated in the movie *Bruce Almighty*. Why can't God do better—we could with magic. Heck, children would make it perfect as was so beautifully illustrated in a poll about this during the movie's ride of fame. Even worse, I put religion to the test of open-mindedness when I asked on the cover of my third book *The Future Alien Contact*, "Could primitive man's God be the universal Roswell alien and not make the world perfect as is the evidence before us?" I know this seemed crazy and dangerous. I wasn't crazy; I wasn't a "sadist"/Satan either. Another funny pun, huh, so I won't mention any more.

But I never cease to be amazed at how religion's creation and molding of our language gives us clues to the mystery of God. Like being awful?

This didn't jive with our standards or their own for that matter. You can't be merciful and at the same time kill, except to end pain. Why couldn't my mom or any religious person get this simple scientific fact! We must be "HUMANE," mustn't we? Again it's simple. You've got it! It's simple. Religion isn't "HUMANE" itself. God is a killer who refuses to make and keep the world perfect!

Religion doesn't teach, anywhere in this world, "THE GOD STORY" scientifically. That's why my previous book, *2012 Gold's History Solves Mankind's Mystery*, utilized good hard evidence, scientifically speaking, to address this issue. But my fifth and final book has a title that directly answers "WHERE IS HEAVEN?" Let's just get down to the brass tacks of it. Let's take it apart—religion, that is. The scientific evidence of God being a real ancient astronaut needing gold for space travel was overwhelming. Erich von Daniken became famous for his ancient astronaut theory but never saw the alien-looking God evidence. The evidence of God

Heaven Is Space... UP!

being "ALMIGHTY" and yet the world being flawed also was overwhelming! The evidence for God detesting wealth but yet living "UP" in Heaven with streets of gold on a golden throne and us kissing his feet was overwhelming. We were mining gold for them from the get-go. This is all so overwhelming unless you open your mind to the possibility that science can explain primitive man's magic spirit miracle story. And ultimately having to accept that religion's omnipotent God doesn't exist! This happened to me! I gave up the traditional RELIGIOUS examination and opted for the "logical" scientific one. Even those two things became synonymous to me. Scientific and logical. They have now graduated from a long "short" process of growth to where they now are today. They have become synonymous in the world of religion, too. Again, read any Jehovah's Witness magazine. They even use notable scientific members to prove that the intelligent design theory of science proves there is a God. They just don't answer why he remains invisible to us—THAT'S THE TRADITIONAL MYSTERY. They don't use science to prove what he looks like either or when he began, just that he's always existed and works through time (HIS MASTER PLAN, WHICH SOUNDS LIKE EVOLUTION) but could instantly make it perfect. Again another contradiction of evidence. It isn't perfect and he has no time, but works through time! Try and scientifically figure that enigma out! Only religion can, and the answer for that is simple. Hopefully, you got it. But if they are using science, it could come back to haunt them if science defies THEIR UNIVERSAL SPIRIT TEACHING or, even worse, God forbid, solve the ultimate riddle. What is God really? What if he's "real" and the answer REFUTES their teachings? Can anything change or cure this persistent disease of religious people not believing the scientific evidence when he comes back? If you're scientific then—you got it—again the answer is no. This mystery exists now. Science rules the universe through "REAL"

Heaven Is Space... UP!

evidence. Logic is believing it. Religion is a story of non-belief when they return. This mystery is incurable. It's all about "Heaven mirroring Earth." We lust for power also. Here, it's all about wealth and beauty/greatness. Maybe we are them, lusting to be greater "looking" for this power.

We all exist now and are alone. This is indisputable and inescapable evidence. This is the "cold hard truth" of our species so inappropriately labeled humans. We are far from humane, and we are "down" on a planet, literally in a universal prison. Since religions don't know where Heaven is and what God looks like then let's look to the scientific evidence for these answers. This was made very clear in a Barbara Walters special. She interviewed every religious scholar across the board and they all said it was a spirit world, dimension, or frame of mind. They all thought that it certainly was crazy to think "UP" in the sky. This is where the simplest of evidence (Occam's Razor) can come back to haunt them. From the beginning we have looked "UP", it has been written "UP", the ancient art like on the front cover and back of this book indicates that it is "UP." Yet they don't see it and most hate me for this. They hate me because I certainly think their "spirit" story is more crazy. At least mine is supported by our spacecraft "UP" today. This matches.

I know their anger is a direct result of having no evidence for their teaching of a spirit world. Thank God I had the ancient statues of aliens and flying saucers. And, yes, I consider my footage of a saucer a miracle, but only in the sense it is rare, scientifically done, and evidence that maybe they are primitive man's God. Heck, we're lucky they even care about us. Wake "UP", people, please? If you can't entertain this notion then please stop religions from saying they are evil. This isn't "right" without the proof to back it up. (The Jehovah's Witnesses do this; see evidence.) And for this reason I set out to answer Fermi's Paradox: why they don't make global and open contact with us. I am doing

Heaven Is Space... UP!

this with scientific evidence. In science the evidence doesn't lie. Knowledge is power. Religion can't stifle this, and why would they—or any of us—want to. Don't we seekers have the same goal, religious or otherwise? Aren't our intentions both good? Could the God story be too "good" to be true? We must accept it if it is.

I can only say that science is a wonderful life, and I love a fast ride in my little red Corvette listening to Prince singing it on the radio. Science made all this possible. So please, my religious daddy, if you exist don't take my T-bird away. Let science be you; please let it be true. This would explain all the hell and misery existing because it can't be stopped. I can only imagine the future of science and the world it will provide for us and our children. We all know this. Not a one of us would deny our children a heart when they need it if we could grow one! And guess what, everybody—great news. WE WILL SOON! This is the wonder of science. Why would religion or anyone be afraid of it. Please, for

MICHAEL

GOD'S SAKE, let us solve the God story. Let not one of us do this if he or she isn't willing to die for our children. This is the greatest love of all, to find something you would give your life for. I can think of nothing more worthy and beautiful for this honor than our children. And indeed it is an honor to be a parent. They are the most wonderfully precious and scientifically logical things to celebrate in this universe. They are the future and they will outlive us and carry on our legacy. Let it be to flourish scientifically.

Religion is killing us with the killing God story. Enough is enough! "Stop the insanity." Will religion let logic solve this mystery by scientifically examining the evidence of our past and its brutality? Or will they continue not to see that "ALL GODS ARE KILLERS"? Do we want this to be true? Do we want to be with "him" if it is? Are we really prepared to get what we wish for if we believe in this "KILLING GOD"? If you're religious, like I was raised, then these questions have to be an-

Heaven Is Space ... UP!

swered because science will do it whether we want "IT" to or not. It is where we came from (increasing scientific knowledge) and it is where we are going. Word ... UP!

If Heaven is Space, as the evidence suggests, then the evidence will tell us how this God mystery came to be and where they are now. But most importantly, it will tell us why they stay away. All of these answers point to Religion 101's first question—or at least it's a no-brainer that it should be—"What does this God look like?" I am willing to accept the scientific or religious evidence. Shouldn't they? Why should either be afraid? I mean, after all, I do believe the scientific evidence of "God" and it hasn't warped my mind. I have scientific evidence to support it. The evidence does convince me that primitive man started the magic factor of religion because of this contact between him and "GOD." But it still has to be answered: What does "He" look like? The little two-lettered, but oh-so-powerful scientific word "UP" yields the greatest clue. Prim-

MICHAEL

itive man's God gave us a scientific fact of life. But primitive man was ignorant of science. This proves what the evidence scientifically confirms. His God was a spaceman! Nobody created the universe. It infinitely re-creates itself over and over again with time, not magic! It exists unseen and bursts onto the scene in an explosion of light to the eye as a gas and then brings forth "touchable" matter. The water of life with all its necessary ingredients becomes aware of itself through evolution. To finally know that it always exists. It is the thing it can't see. It "is" the universe, and didn't come from it. This is infinitely cyclical, just like the ice cube that never stops existing, even though it melts before our eyes and then disappears. And yet we know it still exist as a gas. Why can't we see this about the earth, sun, galaxies and most importantly ourselves!

Carl Sagan said we are the suns/sons of the past and will be them again in the future. What a beautiful reality. They are both beautiful and amazing to look at! I can only conclude and am glad to

Heaven Is Space... UP!

do so, in lieu of my nephew's freak accident, which resulted in his death that I would give my life to have kept his mother and father from experiencing such pain. Wouldn't any father? I saw their pain and it was unbearable! The reason bad things happen to good people is that there is no magic, and there is no reason for it. It is the same reason asteroids destroy life as they collide with planets in a universe full of wonder. Life happens! *Happens* being the key word. Life is magical, but certainly not my nephew's freak death. This was an accident and for this reason life isn't magical because the religion story makes people wonder why God wouldn't stop this. Danny's death was horribly painful, and it helps me to look at the God story scientifically. Religion's account of the Genesis creation mimics scientific evolution, but the magic factor muddles the scientific picture. Anywhere that it is thought or blinked, has to be challenged, like in the very beginning. How can we create atoms or Heaven? There is no proof of this magic process as we know it and science like religion, says it isn't possible. Heaven nor atoms have a beginning or ending. However, "Something does come from nothing," contrary to the scientific Jehovah's Witness testimony of Science proving God is the Intelligent Designer, but yet attributing it to this invisible God. Something/everything is energy/nothing but movement creating an image. I told my brother Brian about this quote and their lack of scientific knowledge. Ultimately, they have to prove their God! If "he" doesn't exist, then so be it, we wouldn't have religious wars! If it came from God and they can't prove what God is then it came from nothing. If it came from him and they can prove what he is then it still came from nothing—he's invisible! Voila!

This controversy still brews, because the contradiction is clear even in their own text. Well, anyway, if they have proof of their God or he's invisible again this means it came from nothing! If we are able to deal with this fact, then life is still beautiful. Except for one thing, my precious nephew's tragic

Heaven Is Space... UP!

accidental death, as well as my little sister's, and every other child who is a victim of it! For God's sake, people, their's, and ours—there can't be a logical reason for freak tragedies.

Please let us solve the God mystery and our own? I propose that they are maybe one and the same. Could primitive man's god be a real flesh and blood alien?

Please check out my upcoming book...

The Discovery

MICHAEL

Are all the ancient religious monuments made of huge stone circular patterns, like Stonehenge, really just flying saucer representations/art?

Heaven Is Space... UP!

**December 23, 2012, "The final frontier is 'BAK'!" The story continues.
(What goes around comes around? Life is indeed an atom, yah!)
The mayans believed in infinity
it's all about the money
The Golden Age
RETURNS
NOW,
UP
!**

"All right, everybody, it's happening," I said as I looked "UP"! I looked over at the President and he was looking up, too! *Wow*, I thought for a split-second as a big smile came over my face. Everybody must be looking up! Finally, science was making religious people come face to face with this "UP"/space-traveling/flying saucer reality. Without religion we'd be expecting contact from space! Why did it take us so long to realize this? But have we really?

Why does most of the world not think this way? I knew why: religion. But today, science won out! The president was religious and yet he WAS looking up! If only he could be open-minded to his Jesus being an alien, which looks like all the other aliens. One species whose worst nightmare couldn't be stopped: "HUMAN CREATION."

This was Heaven's occupant's lust for power story in religion, "THE FALL." Our unique looks give us power over one another. We're all addicted to beauty of the flesh. I knew it all too well. Hell, just watch television. Beautiful women and fabulous wealth is the ultimate reward here. This is as much an irrefutable fact as heaven's "UP"-ness!

It was D-Day for me as well. Was I cured? Was I really ready for this ultimate reality of mankind? The whole Earth must be looking up! The saucers began to land. I watched as they descended within seconds. I couldn't believe what was happening. I was just as excited now as the first time I saw one. My mind was racing with all kinds of thoughts. What had happened to me in the past six years was incredible. I finally saw and filmed flying saucers.

But enough of this, I thought to myself as I

Heaven Is Space... UP!

turned again toward the president. I had to stay focused on what was happening. I was worried about little Jake. I knew little Jake was okay, though. He loved riding on a saucer. I had taken him on one after my arrival back from Jerusalem. And I sure wasn't about to stop him from coming today.

Besides, he wasn't little Jake anymore. Although, I knew he'd always be little Jake to me. My anxiety left me as my smile grew even wider. I knew he was looking down at me now. Nothing would do but for him to come with me this morning. *Little Jake*, I mused. I laughed aloud and maybe even cried tears of joy. He may have grown up. But just like I was always little Mikey to my dad and older brothers, he'll always be little Jake to me. Somehow, I now felt good about all this. I knew deep down that he and Jimmy were all right with it. Hell, it was Angel and the rest of this crazy-ass world that I was worried about. Especially the religious ones.

But then again, I knew they would be stopped if they resisted. Why did I know this? I guess it was just scientific logic that dominated my thinking. I was always amazed at the SETI program thinking it would make contact with extraterrestrials. Everything scientifically and logically pointed to them being more intelligent. It was ridiculous for me to think beings who have already conquered space would respond to our repeated attempts to make contact. There wouldn't be a mystery if this was the case. Believe me, if a more intelligent species could help us, they would! This is only logical intelligently speaking. The same thing applied to the God story; he loves us and created us because were a "good" thing. If we were either "good" or worthy of being contacted then this mystery wouldn't exist. Their reason for not bothering is clear by our disease of war and lust for power over one another. We are the most destructive thing in the universe. It's unbelievable that people don't think scientifically about this lack of contact from an intelligent, loving extraterrestrial. For god's sake all our wars are over religion. Why wouldn't God or this extraterres-

Heaven Is Space... UP!

trial just never let this mystery began. Hell's bells, people, they could just stay up in the sky and never let us forget why they don't live with us. There must be a logical reason they don't. Serving a purpose working for them is the only thing that makes sense. How did religion make us a good thing and how does it continue to hide the provable fact that they live in the sky as real flesh and blood beings capable of death, pain, and addiction just like us? When religious people would counter—and they always did—that god did this because he wouldn't want robots, it made me sick. I'd always tell them how it would stop war if he would just make it clear what the hell he is!

What hogwash and lack of logic, not to mention cruelty. It's illogical and down-right sick for any loving parent or adult to let innocent children be tortured and killed to prove our "LOYALTY" to this "mysterious" God. What a sick compromise from people who wouldn't do the same nor would want to be worshipped either because society sees it as a sickness. I can't stand their free-will argument. Hell, the word wasn't even in the Bible. And it still doesn't take away from their God being responsible for all mankind's hell and a chaotic universe. Why don't they see he is a killer of innocent children in the Sodom and Gomorrah story? Why?

If all this isn't bad enough, they have another creation gone awry to add to mankind's temptation, and that is the devil story. Their God is inept and a killer to boot. Besides being a sadist who chooses not to make it perfect! In every book I drove this PROVABLE POINT home! Come on, religious people! Where's your logic, when you demand it from your children? This religious sickness was almost just as bad, though, as scientists who think this lack of contact is due to the vast distances in space. Isn't it logical to think that the infinite universe is already conquered? Concepts like distance and time are irrelevant because they wouldn't live on planets. They would just rest on them. I always quoted Yeshua/"Jesus" for proof of this reality.

Heaven Is Space... UP!

He said "HEAVEN IS THEIR THRONE AND THE EARTH IS A FOOTSTOOL!" They also don't have time or distance! Isn't it logical that if we have "UNIVERSAL" stories of beings that live in the sky, to look for the answer as to why they stay away? Shouldn't we be realizing that these flying saucers must be them? I knew why religion didn't think this way. My search is about heaven's change in status from "UP" to a spirit world.

But isn't it logical to never let religion make heaven some place other than what the facts clearly state? It is "UP"/sky! Why doesn't science see this "UP" word as "CLEAR" proof that space is conquered? How can we have scientists who believe in spirits without proof? Isn't it logical that this one common theme is enough to make us all see this religious being/god is real and not a spirit magic guy who has already conquered space time and infinity! I suddenly realized that, in my barrage of scientific wonder and horror, the most powerful man on Earth who believes in this very thing was standing beside me looking "UP." Now, even he was faced with this harsh reality. The seeing-is-believing factor made us all swallow this cold, hard fact. It stood before us, front and center.

The flying saucers were landing now! I looked at the president in total disbelief. This was happening! THE PRESIDENT IS LOOKING UP. We all were! Again, I had a sudden chilling realization that I had not only seen one way before this, but had actually filmed them. What! *No way*, I thought to myself. *This can't be real*. Wow, but it was.

It was all coming back to me now. I did do this. I had done it with a man named Jeff Willies from Phoenix, Arizona, back in 2005. Seven long, hard years ago! Sure, my books got me world-wide attention and hatred. But this event became my final obsession. This piece of evidence became the catalyst for me to find an even bigger piece of evidence. With it we would become *The Jeff and Mike Show*, *REAL-LIFE FLYING SAUCER HUNTERS*!

Before I go on with CONTACT, let's first get

Heaven Is Space... UP!

back to how our paths became destined to cross in the first place. And how fate teamed us up to find and film REAL-LIVE FLYING SAUCERS. And find them we did! We're still finding them now! You'd think that after actually doing this six years ago, it wouldn't have taken so long for us to garner world-wide attention. But it did. There were lots of photographs of flying saucers, Jeff's included, and believe me, it along with mine were big. I'll tell you in a second why it was so big, but for those of you who have already read my books, you know!

But, until we proved it by showing them live, we were still ignored. And then we couldn't be ignored. Television time could be bought, and we bought it. Throughout all of this amazing evidence that I had put in my books—and I do mean some truly amazing stuff—none of it worked. Not the evidence of Jesus being a false name or the criminal changing of it, to finding a great saucer picture. And not just any picture. But a picture of a saucer that had none other than one of the most famous entertainers of all time in it. HE WAS ALSO LOOKING UP AT IT!

I had found a photo of Johnny Cash that matched Jeff's flying saucer footage from the infamous Phoenix lights. I used it in my third book. It didn't get the response I wrote about until . . . read on. Well, that picture is real, but the story part was fiction. Like I said, it didn't get me world-wide fame. Hell, it didn't even get the attention of Nashville and I protested in front of their newspaper! One of the editors came out and told me he couldn't review every book. He said there were plenty of pictures besides mine. I told him, "Not with the Man in Black." I quickly added, "Get it, 'man in black'!" He wasn't amused nor impressed and seemed a little angry. I asked him if he believed in God. He said yes. I knew why he was angry!

That's why I finally use my real name on this last book. I knew it was real; this wasn't fiction. But hardly anybody thought that me putting the evidence together would solve our mystery. Thank

Heaven Is Space... UP!

God for my brother Brian and Terry! They finally would. Seeing is believing to them as well! Well, I'll get to that in a second.

First, let me tell you what happened as a result of finding and using this picture. This is the actual story. This is what really happened with no embellishments. In 2003 when I found "THE PHOTO" Jeff and I became great friends and immediately set up a venture together. This resulted from an attempt of mine to get on a nationally famous radio program that was using Jeff's photo. Like I said, I wasn't famous; that was fiction. But this event was real. My fame wasn't world-wide, anyway. I, well, we would soon get there. I had no clue that he would become the new "other witness."

Anyway, if you read my first books, you'll know what that reference was. My first book is called *The Two Witnesses and the Religion Cover-up*. This will give you a clue. He was religious just like my other buddy from the beginning book. But at least he had found and believed in flying saucers. He still

MICHAEL

believed in spirits, though, just like my other buddy, and he didn't have any answers for why they wouldn't openly live with us, either. Spirits can't be hurt or killed. Well, he did have the stereotypical ones "like maybe we're not ready yet" or, like I said before, "God wouldn't want robots." These were both illogical if "he" loves us. To make something lovingly perfect through pain is ridiculous. Not to mention the freak accidents and senseless acts of killing and torture! But they both had the worldwide religious answer, too! It included them to be demons, and I sure couldn't stomach that lack of proof or logic.

It was, in fact, harmful to our purpose of wanting to figure this mystery out. It blinded them to scientific logic. It scared the majority of people. It was terrorism! I showed them both that they were no different than the Jehovah's Witnesses. They refused to see this or couldn't! I thought these were one and the same, and then I saw a similarity in science. They don't see the "UP" significance of reli-

Heaven Is Space... UP!

gions heaven. Maybe they didn't see it, even when presented matching evidence, science didn't. I didn't know how this was possible, except for their stubbornness to change theories. I always liked what the show *CSI:* and Grissom would say: "If the evidence changes so must your theory!" Theirs didn't, just like the rest of religions.

Anyway, I presented this evidence in my last book as well to prove it to them. They saw the matching evidence and refused to change their belief in spirits! I had used proof of this religious reality in my last book. It was universal. It was truly a religion cover-"UP" on a universal scale.

Back to the radio show and possible interview. Here's what happened. They declined an interview with me, but told me my picture looked similar to another famous photo and then used Jeff's to compare. This is how I met Jeff. I finally asked the host if he was religious, too, and he said yes! He was a Christian. Wow, what a struggle. Scientific evidence vs. religious pride.

MICHAEL

Anyway, this one rejection started the *Jeff and Mike Show* rolling, whether I knew it or not. I didn't. Believe me, I too, never thought I'd see a flying saucer, either. It became an adventure that would change my life forever. It really was the catalyst for our fame—and for one huge reason. I believed Jeff, and we found and filmed saucers! I first saw one that year before we did. That blew me away. I didn't get a picture of it, though. But when Jeff and I filmed them and got on television a year later, it received world-wide media recognition of flying saucers, then and now! It wasn't just this, though. After all, I had a picture of Johnny Cash and a flying saucer, for God's sake. Why wouldn't it? Right.

Well, seeing is believing, believe me! Anyway, I was using this as proof that aliens in flying saucers started religion. I had compared it to a cave drawing of one in France on the front cover. We even made billboards together. (I will write about this later. It is what happened in the years succeeding the third and fourth books. All of this was eeri-

Heaven Is Space... UP!

ly coincidental. It all led up to and precipitated this world-wide contact by flying saucers. Remember, this is 2012. I have déjà vu, and religion knows our ending. It's from the sky! It is now. Coincidence or Destiny?)

Well, I used his video picture again on the fourth book and had advertised a video using this footage called *PROOF*! But it was Jeff's and not mine. Remember, I had only seen one in the fall of 2005 and I don't think anybody but him truly believed me. He did and, needless to say, I believed him one hundred percent! And now I was going to film them. At least I hoped. No, somehow I just knew it. I saw one. I was so excited. I hadn't filmed them yet. But, I had pre-advertised a new documentary. Little did I know what was about to happen, and I just couldn't pass this off as a coincidence. It would be the smoking gun of all smoking guns! I was going to compare Jeff's footage to the world's largest ancient statue of one on top of the heads of Easter Island and many other ancient pieces of evidence.

It turned into something much more. He had told me in 2005 that I could find them with him. I wrote about his evidence and claim to find and film them in my fourth book. This claim drove me to investigate after participating in the International Film Festival/UFO conference. I went straight to his house afterward. Before I left, though, I have to say that his proposal to find flying saucers was pretty unbelievable to me. Even seeing one wasn't enough because I hadn't seen any more. Besides the fact that I couldn't really afford to keep trying!

But something happened that drove me to give it a chance. While I was there, I had a contact moment which left a mark on my face. It was filmed by a local TV station there in Laughlin, Nevada. Jeff also experienced contact. Needless to say, I went. Nothing was going to stop me now! When I arrived at his house, we immediately set up the gear outside and started to relax for an evening of sky-watching. I told him that it was funny for a guy who believed

Heaven Is Space... UP!

in spirits to be sky-watching. He didn't comment. You see, I had asked why we were sky-watching back in Nevada, if everybody thought aliens were among us and believed in spirits. I reminded them that Heaven's UPness was irrefutable, like our going up to conquer space.

Nobody cared to comment. They were tired of me. They believed heaven was a spirit world or didn't know where it was. Neither one made any sense. None of them said UP. Why? Why didn't they get it? I knew why! But why didn't they look for spirit proof? The out and out lack of hurting or killing a spirit begged for logic besides this OBVIOUS MATCHING EVIDENCE! Why just an apparition or white noise, anyway? They can't be hurt if they're a spirit, can they?

I couldn't get them to see the lack of logic in this. I couldn't get them to understand that this doesn't reflect evidence of a spirit being, but one that is real and can be hurt or killed! Worst yet, I couldn't get them to see mankind as an evil cre-

MICHAEL

ation, and this is why they don't help us. We can't be helped, that's what the religious story says! The majority is predestined for doom. And God forbid that I tell them that we could be an alien addicted to the scientific possibility of being in another body that we created in a lab. One that is used for its outward beauty and its powerful effect over other humans. One that we can control or at least influence its thoughts.

This came true in 2006. We finally achieved "scientific" invisibility and made robots that we could control with mental telepathy! Nothing I said or showed them seem to work. Not even their own admission of wanting power and being addicted to outward beauty, MATCHING the religious story!

Anyway, that night I found them! Before I go on, I want to make it clear why I used Jeff's photo from 2003 again on this book cover. It's because I videoed the same one at his house in 2006. The next day! Yes, I did it. Look at the picture of it compared to his and, yes, Johnny Cash's. I couldn't believe my

Heaven Is Space... UP!

eyes, and it solidified my futuristic scientific thinking that space was conquered. By all appearances they exhibited all the characteristics of HEAVEN'S occupants. They could cloak their craft and render themselves invisible. Their craft had no sound! They weren't hurting us, nor helping us either. They weren't stopping accidents. Therefore they couldn't predict our every move. However, they did seem to know that we were watching them. Mental telepathy seemed to be obvious as well. I watched them for three days and could hardly bring myself to leave. I think they knew it! But again, they know we're an evil creation. We just don't know it yet.

Why help us to prosper? Maybe this is the way it is. Us wanting to be them isn't forcible, nor possible. We're just a disease running its course. No, we're gold-diggers and they need gold for speacecraft. Our every move can't be predicted, just our/mankind's outcome. The best proof of this is Yeshua/"Jesus" not knowing the guard's ear was going to be cut off by an apostle. I mean, come on. It

MICHAEL

doesn't take a rocket scientist to figure this out. And what about him asking why his father had forsaken him? This doesn't make sense if he already knew he wouldn't.

I always wrestled with this religious teaching of pre-destination, but worse with my own déjà vu. Déjà vu indicates we've done this before. Religion and scientific probability points to this reality as well. The idea that energy can't be created or destroyed and that all matter is energy, agrees also. But these scientific concepts are "deep," like infinity itself. I thought all of this wasn't as important as contact. We have a history of alien/flying saucer evidence, and I filmed one. To me, contact would eliminate our doubt about everything. It does! Read on!

Our doubt about not existing is really selfish. Why else do we want to know if the atoms of our body will always exist? Knowing we are atoms/energy isn't it ridiculous to ask, "Could they ever be manifested as me again?" If we only had one shot

Heaven Is Space... UP!

wouldn't it be best to try and conquer death for our children? How could we be so selfish and expose them to the same fear of death that we have? The sad reality of mankind is that the majority of us think we would be brutal without religion's reward-or-punishment doctrine. The evidence indicates otherwise. More deaths occur from religious acts of violence than any other. This is second only to people committing crimes for money, but proportionately quite larger. I hope this information provokes research into such a startling fact. Are we truly better off believing in life after death? Wouldn't that make us even more likely to live viciously, for the moment thinking we will have another chance? Maybe, we would have wealth, beauty, and power next time. The evidence indicates we are brutally and religiously vicious. Does the killing god story create this? Please consider it. He is a killer.

I was recently studying the scientific research of Michio Kaku. He stated that the universe is mostly composed of much more than atoms. In fact

MICHAEL

the majority is made up of dark matter (fragmented atoms). It is made up of elementary particles. I dug and dug until I realized that the smallest part of matter is termed nihilistic atomism. It is really all that exists. However, it is moving and only a part of the whole which is constantly appearing and disappearing. This is microcosm science and deep, but I understand it from the ancients. Almost everybody knows the quote "I think therefore I am"! This ancient quote is clearly quantum physics. But I haven't found too many people who know what an atom is, let alone, it being what they are made of!

So, it is simpler for me to use the sparkler at night example. It will explain how this "movement" relates to the infinite nature of matter, just changing form. The circle you make in rotating it really doesn't exist, just the sparkler. We are the sparkler/energy that just changes form! So do we, the universe, or anything the atom makes really exist? In effect we don't exist permanently, but just as a temporary image. But we need to remember we are the

Heaven Is Space... UP!

"moving"/stationery atom. I like the ice cube example to explain how we don't doubt the infinite existence of gas and its ability to turn into water then freeze and ultimately disappear. Yet we know it still exists.

This is the holy grail of mankind to understand that everything is nothing but moving energy. it didn't come from anywhere and isn't going anywhere. It is the godism quote of the ancients and Michio Kaku's theory. I wanted to take the time now, before I go on to share my latest discovery as a result of looking at Michio's umbilical bubble theory of the universe. In a moment, I will return immediately to March 6, 7, and 8, the date that I first FILMED FLYING SAUCERS!

Anyway, back to my overwhelming scientific discovery. I was shocked, after watching an interview of Kaku recently on *HARDtalk*, to find him answering questions about the universe. Like where did it come from and where is it going. He was clear about it being a cyclical dance of expansion/birth and contraction/death. He was also clear that we seemed doomed by it. Wow. I listened on—he was talking my language. I wanted to see if he saw the parallel to Buddhism and nirvana. Their goal is to break the cycle of birth, death, and reincarnation. If the universe has us doomed, religion has us doomed, and science sees the likelihood of our doom, given the ability to destroy ourselves with nuclear weapons, why bother with all of this? He said that it is our nature, and I agree! We are in effect gods of science, and it is in effect what drives him to try and overcome our "mortal" challenge.

He sees this as an exciting time to discover the tantalizing possibility of a loophole and escape the expansion or contraction of these bubble universes (the multiverse, which is really "one" infinite universe made up of an infinite birth of new ones). He sees this opportunity, through what appears to be a white hole umbilical cord blowing out of a black hole. Talk about blowing out, I was blown away! I saw this very solution to achieve the immortali-

Heaven Is Space... UP!

ty of our species several years back and put it in my fourth book. However, my hypothesis was based on the membrane theory. It says that space is flat, which makes sense if it is spinning matter out from a central point. I felt the ability to stop this cycle, birth and death of a universe or membrane of matter/universe, was only possible through conquering space. We would have to out-distance the gravity field caused by it. My theory is based on space being flat as supported by the cave grid. His is made of bubbles, which I think could also be flat.

Either way, our principle theories are the same! He went on to say that science gives us an edge over the inevitable destruction of every universe, which is just one of infinity. He said we can only do this through space travel! Wow! My point exactly! If I can just show him flying saucers, maybe he will get my matching "UP" point. Religion is a story of people who *live* "UP" in the sky! I say the ancient evidence will give us the answer that we are so desperately looking for. Like the cave art of a grid showing our solar system exactly with a rocket/ shuttle and the asteroid belt! It will prove religion was created by space travelers. I wanted to show him that his bubble theory is supported by the ancient religious yin and yang model. It has a black and white hole, the spinning motion of both galaxy or black hole, and looks like a bubble. It also has two sperm in an egg, which maintains life through an umbilical cord. I first used it to confirm the story of angels/ gods mixing with man, as two sperm in one egg would do this. But, now I see it could do both—and indeed does. Whoa, the parallel of religion's message is the same as his goal, to achieve the immortality of our species! Religion is a story where they are omnipotent, omnipresent, and omniscient. They have conquered space and live in it, just like he and I theorize that we could do as well. We would do it by going "UP," and they came from "UP!"

Wow! I hope he sees this MATCHING EVIDENCE. It irrefutably proves space is indeed con-

Heaven Is Space... UP!

quered. The matching of ancient religious symbols to today's scientific ones are amazing and universal. If this doesn't convince him then I will show him a flying saucer. This will do it if he's a "true" scientist, one that follows matching evidence. If it does, then maybe I will get some respect for my scientific theories. First, they have to be watching us! We both agree that life in space has to be infinitely more advanced than we are. We also both agree on the natural progression of this intelligent being and that is to live in space. If I can continue to film and find flying saucers, then it will make us address Fermi's paradox. Answering "why they don't contact us" with logic will force us to accept this ancient religious evidence. They have to be real flesh and blood people who live in space, just like they all say. And they "CAN BE KILLED JUST LIKE US."

Secondly, they can't cure our primitive nature. The final question to be asked is why make a destructive scientific creation and yet we have already

MICHAEL

answered that with nuclear technologies! It can't be stopped, serves a purpose, and creates the desire for power that it will afford us! WOW! This is our story, people. Finally, a eureka moment that gives me a simple understanding of the universe I live in. It is infinite and space travel is the only way for scientists to separate themselves from the brutality of mankind. Thank You, Michio Kaku!

I went through all this hell the past fourteen years only to confirm what I always felt inside, anyway. It was my CON"SCIENCE"! Wow, we can't "CON SCIENCE." Man, these word coincidences are a trip. What goes around comes around. We can't stop existing. We are atoms/adams! I really hoped to explain the structure of the universe in a much simpler way. The big bang starts with the atom, and in Buddhism we must become one with the self. Adam must learn he is the atom! Wow, I discovered this to be the ultimate scientific knowledge that children and adults lack, myself included until now.

Heaven Is Space... UP!

Asking what the universe is became an eye-opening question for me. I asked Angel, and she said it was everything as we know it. I asked her what is everything. She said it was us, the planets, space, etc. I asked her if she thought there was more than one universe. She said, "Of course, I just heard of the multiverse thing that you were talking about the other day."

"Well," I asked, "Where are the other ones?"

She didn't know. I asked if they were next to ours. She didn't know. I asked how there could be multiverses if there was only one universe. She didn't know and looked really confused. Hell, this was understandable—almost everybody is including myself. I wanted to ask her if she is the universe, but know better. I always sang John Lennon's *The Walrus* song, where he sings, "I am the eggman, I am the walrus, koo koo kachoo!" We are the universe. I wanted to explain how giving numbers to anything infinite is an oxymoron. I could see that Michio was saying that the multiverse theory is really infinite. Therefore, the term *multi* is a contradiction. To better clear up this confusion over one universe being composed of infinite multiverses (again, this is an oxymoron: one being multiple vs. infinite) it would be better to call them fields of matter as stated by the membrane theory. The reason I say this is that they are not really separate from the others. The gravitational pull is the only thing that separates them. The black space is infinite and expanding through the creation of baby "universes" born out of the bubbles of matter.

This would agree with both scientific theories. It also confirms the way matter continues to be born out of the union between it and anti-matter. This intercourse leaves one part matter over the one hundred per one hundred parts explosion theory. Therefore, this is what creates and sustains the ongoing and infinite existence of the universe, which is really just matter. We are it. This is so, because if it didn't exist, would any of this really matter?

Heaven Is Space... UP!

Therefore, we have now come full circle to repeat an ancient religious philosophy of "mind over matter" that has become today's science. To conquer our understanding of it is secondary to space travel. Funny that it might take actually conquering the atom and making anti-matter to achieve this! We are doing this with the particle accelerator, but at a very slow rate. I was surprised at Michio stating that another universe could exist within his living room. I was always under the impression that two things couldn't occupy the same space, like matter and anti-matter. I will look up this law in the deep theories section of my physics book. I find it hard to understand how another "universe" could fit in such a small space given the expanse of what we "think" ours to be.

Anyway, I think that calling these flat fields of matter floating in an infinite universe a "universe" or dimension itself is a mistake. It gives the person the impression that is world-wide today, dimensions exist. This has led people to speculate that worlds/

universes can exist within matter itself. I think this is called the "grandfather paradox" of time travel. The bottom line is that if there were other dimensions we wouldn't have the "UP" evidence of these religious beings and we wouldn't be going "UP." Spirits nor dimensions have yet to be proven and until they are, the evidence should dictate which is correct. We are going "UP" and have ancient stories of people who live "UP." I film the sky/"UP."

Now, back to my flying saucer career. It's still hard for me to believe that I not only saw one in 2005 but actually filmed them this year. And what's really hard is having to finish this book, before I can chase saucers again! I know I can find them now! Let's go see what happened. I'll give you a hint: nothing! "Just kidding," my Uncle Brice would say. Well, at least nothing had happened yet, anyway. Read on to find out if it does! I will tell you though, if it did you wouldn't be reading this book and not know about me.

Remember, up until now my story of world-

Heaven Is Space... UP!

wide fame was fiction, excluding the scientific discoveries. Now, I am telling you the real ending! If you look at the copyright date on this book, it says 2007. It is right now Christmas of 2006. I will be releasing this book in February 2007. It's been almost a year since the Barbara Walters special was on. While I was typing this it was advertised to come on again. I got pretty excited about this. Maybe, I would have another chance at getting her attention with this new book and new video. Here's why. You see, I gave her my fourth book and video proof last year, right after I left Jeff's. Well I gave her *WHY THE BLANK DON'T THEY CARE*, too. But, it was mainly just the footage I told you about already. I will get back to that in a second, though.

But before I do, I know now why it didn't get her attention the first time. Maybe it's fate. I mean, after all, my last book did say "SCIENCE FINDS HEAVEN," but it didn't "prove" it. I think it was the combination of a poor title (*2012 Gold's History Solves Mankind's Mystery*), no still shots, and poor

MICHAEL

video. Now I will have a new book and a new reality-show quality video with amazing stills! And this book has a title that "CLEARLY" addresses her question: "WHERE IS HEAVEN?" Besides, I'm better prepared. I got several still shots of my video saucer footage, daytime and night, to prove that's what they are. The footage was shaky, but the pictures left no doubt about them being saucers! This new *WHY THE BLANK DON'T THEY CARE* video has a new beginning that also pitches us as "THE JEFF AND MIKE SHOW: REAL-LIFE FLYING SAUCER HUNTERS!" I hope to attract the investigative side of her television nature and capitalize on the reality show phenomenon that's so hot right now. To me this is the ultimate reality. Entertainment would take a back seat to "NEWS" if we could do this. Heck, the whole world would be immediately drawn to catching one or making contact. They would converge upon Phoenix and other areas of high activity. And wouldn't we want to know why just these spots and not the same every-

Heaven Is Space... UP!

where? Are these places like Phoenix and Mexico City in jeopardy? This would take first priority all over the Earth. I am sure of that. I'd tell you why I'm sure, but that's my ending. I know we can find them, and we will! You'll see why it becomes the "SMOKING GUN' as I told you earlier. We did it once; we can do it again!

So follow me as I go back to where I left off that fateful day—the day I actually filmed saucers in Phoenix, Arizona, in 2006 with Jeff Willes. Wow, that still blows me away! Now, let's continue on: here's how it went. I left off at Jeff's house in Phoenix, March 2006. I had one video out called *Proof* with his famous video shot from 2003. I had put it in my fourth book and was already advertising my new book, the one you are reading. Coincidently enough, I had also advertised a new video called *WHY THE BLANK DON'T THEY CARE*. But little did I know I would find them as I headed out west, that fateful cold February morning before all this happened. I wish I could have filmed

MICHAEL

myself then. As soon as I pulled out, I began talking about how crazy this was. Did I really think that I could just drive out there and find them? I literally laughed out loud at the thought of this coming true. And believe me, I did talk to myself, a lot! It was crazy, or I must be. I couldn't stop thinking about it as I drove, because I hadn't found them again since I saw one over my house in 2005. And I looked hard damn near every day! Since I hadn't been able to find one for over a year, it seemed ridiculous to think that I would. I was losing hope, and I certainly never thought, for one minute, that Jeff and I were destined to do this. But we did!

Hell, I was naive enough the first time to think my buddy Jake and I would find spirits or even Yeshua/"JESUS" himself. Jake just knew he was a spirit! Right. He never proved it! As a matter of fact, the few times we experienced the apparitions or a voice out of thin air, it made me even more certain these angels/gods were real/flesh people, and this was proof of that. This is how real advanced

Heaven Is Space... UP!

people who will be hurt by us would communicate. Anyway, we never made contact/open communication, and they never answered questions telling us they were spirits! This to me was what we all were looking for, like the Ghostbusters. A spirit is the "HOLY GRAIL." Well, like I told you earlier, he wasn't looking for them scientifically. I became the Ghostbuster. He is a spiritualist!

Man, what a turnaround over the past fourteen years. I went from believing in spirits to ACCEPTING THE EVIDENCE OF ANCIENT ART DEPICTING ALIENS AND FLYING SAUCERS! My theory was now based on "PHYSICAL EVIDENCE"! Surely, advanced technologies of these aliens in flying saucers must've created the magic spirit factor of religion and its universal stories—incredible scientific stories of fiery chariots taking people (which sounds like abduction) "UP" in the sky. It was this ancient evidence that assured my sanity when I did see one. It was "UP" in the sky as well. It was then that I was one hundred percent

MICHAEL

sure they had seen them, too! Now it all made so much sense to me, especially the hell of life. They were "real" flesh-and-blood beings and not spirits. Now we know what's real and what isn't.

Funny, I remember as a child how Mom would tell me that spirits didn't exist. She would do this to calm me down after I got too scared in church or after one of Dad's sermons at home. The irony is that she really thought they did. They scared her, too. (Check this evidence out; it is both sad and ironic.) They made her scared of the very thing she was supposed to be looking for! If she only knew from the beginning that it was her belief in something she couldn't prove that scared us in the first place. But she never proved it, just like Jake and my new "OTHER WITNESS." The whole world began to hate my challenge to prove a spirit. But they wanted me to prove the saucers. I do! No, we do! First me and Jeff, then the whole world.

Oops, I am telling you a little too much. Remember, it hasn't happened yet. But earlier I had

Heaven Is Space... UP!

said the world might be better without religion. Well, at least we wouldn't have this fear of monsters and spirits. We wouldn't have religious wars either, huh. That's a given. Wow, what an unbelievable transition, huh!

Well, anyway, back to finding saucers! So, like I said before, after my sighting of 2005, I looked and looked but never saw another one. And in not finding any more, I had already planned to use more of Jeff's footage and ancient matching evidence to support my theory. I was going to prove that flying saucers had created our religions. But what a surprise I got, huh! I filmed flying saucers! Just imagine me doing this with you. Pretty crazy, huh? Well it gets even more eerie. I told you about us making billboards. Well, I mean literally. You'll never guess what happened, or maybe you will.

Anyway, I'll go ahead and finish the story up to the present. It's Christmas of 2006. I left the story leaving Jeff's house after three days of filming over ten flying saucers. I called everybody I knew—fam-

MICHAEL

ily, friends, and some potential newspeople, thinking that would definitely seal the deal on my fame. But it didn't. We got on TV, however, our short three minutes of exposure on a local public access channel failed. We couldn't believe it.

I left Jeff's telling him to get ready and board an airplane to L.A. or New York. That's where they were telling me to go. I did. I left Jeff's and went straight home to prepare for this trip. I now had footage to make a new video. This changed my whole plan for *Why The Blank Don't They Care*! I wanted to show the whole world that the media didn't jump on this! I needed the world to see how "OUR" saucers matched the hats of Easter Island and that this was not only an ancient statue of a flying saucer, but proved religion's heaven was "UP" and caused by flying saucers. So, that's what I started doing the days, weeks, and months upon my return. I found it harder and harder to be normal again. I really became the weirdo now. I did, however, find more saucers over Nashville and tried to

Heaven Is Space... UP!

show one to my brother Brian, but he didn't see it. I never saw any more over my home; even as I write, I am looking. But I did when I started building my billboard. That didn't happen until early September. In the meantime, Jeff finally got our new video done without the new reality show introduction. He told me that would take awhile. We were doing this on a shoestring budget. Man, it was just killing me to wait. My book sales were down to nothing and even my trip to Roswell didn't do much good. I always do their annual Fourth of July UFO festival. While I was there I advertised on the street: GUIDED FLYING SAUCER TOURS! Nobody bit. I couldn't wait any longer. So, I went to New York with a three-foot poster showing Jeff's saucer over the Easter Island statue's head and my new unfinished video. Naturally I didn't have this book yet, either. But I was hell-bent to show Matt Lauer my new Easter Island evidence and tell Barbara Walters that I did know heaven is "UP." I thought my poster would prove it! It didn't.

MICHAEL

When I returned I became very depressed. I called all the TV stations in Nashville and never got one bite. This drove me to do a protest in front of the newspaper, located right downtown on Broadway. I got a lot of support from passers-by, but still no press. All I could think of next was the billboard. My brother Terry lives on a major four-lane highway in Ohio. I knew it wasn't going to get me the attention that I could get if I jumped the gates of the *Today* show. I actually thought about this. After all, I had live footage of flying saucers, for god's sake. Needless to say again, I didn't or you would have known about me! I went home. The protest thing is what spawned the billboard idea. I had a lot of people come by me. This proved to me that patience is a virtue. I wasn't going to jail when I couldn't change things anyway. I wasn't able to just make them show up. Hell, I couldn't even spot any more, after that. How could I with everything I had going on?

I needed to start this book. I knew I would

Heaven Is Space ... UP!

have to have all this in place, anyway. I desperately needed the money to do both—the book and a full-time flying saucer hunting career. It wasn't easy, making the money and finding more flying saucers. But in the back of my mind I knew I could guarantee a sighting in Phoenix the first day. The amount of activity there vs. Nashville is phenomenal. However, I did still see one after Brian left in November. I did it on the way to Nashville while I was making deliveries for business and seeing my editors. They were pretty excited about it. I wanted to put them in here as witnesses to the event. Even better though was finally getting to show one not only to Brian but my sister-in-law Vicki too! Brian helped build the billboard in September. The first day after it went up, I spotted them as he was getting up on the scaffolding to help me stabilize it. It was twenty-six feet high! I couldn't believe it, but as I looked "UP" to guide him, I was blinded by the sun. I quickly moved over and let the billboard block it. That's when I saw them! I saw one big one

MICHAEL

and a faint smaller one off to the side just above it. I really got excited and started screaming for Brian to look at it. He thought I was bullshitting him, but I yelled, "I'm not, goddamn it, just look." After not being able to find it on the scaffolding, I didn't think he would believe me. But he did, finally. How could he not? I was screaming and telling him I saw another one and . . . and.

Wow, I saw two definitely and what appeared to be some flying by it. A lot of them really high "UP." They were really visible in the white glare that I was looking at in blocking the billboard. He had a hard time finding them, but when he did it was great because he didn't just dismiss it. In fact, he abruptly replied quite surprisingly, "What the hell is that thing?"

"It's a flying saucer," I remarked. (I remember Jeff telling me the same thing back in Phoenix, when I first saw an irrefutable UFO I got on tape.) I hollered again that there was more than one.

He said, "It ain't either, goddamn it."

Heaven Is Space... UP!

"Well, what the hell is it then, if it ain't?" I snapped right back.

"I don't know, Michael Duane," he said, less angry this time, "but you've got an awful big imagination." He called me by both first and middle names like mom or dad did when they were angry. I laughed.

"Well, what is it then?" I demanded. "It is round and it ain't no balloon, that's for damn sure. Now is it?" I really hate people doubting me like I'm a big liar.

"I don't know, but it ain't a flying saucer," he maintained again. "Well, let's agree upon what it isn't then. But first let's just look at the matching geometrical shape of it and a saucer. It does look just like a round saucer that you put a cup on, now doesn't it? Let's also look at its color. It's silver, shiny, and spinning, right?" I fired all these questions at him before he could respond.

He really started looking hard and made it a point to challenge them all, but couldn't. He did agree that it was round, shiny, and moving. He started to say it was small until a commercial jet flew under it. I got his response at this point to use later on as a scientific witness who was challenging my claim. It was beautiful. Because after challenging everything else—shape, color, movement, etc.—he was forced to examine the last thing that would rule it out, its size.

I screamed as the airplane came toward it and said, "You see, there's your proof. It's huge." It was twice the size and he used the comparison of it being a fifty cent piece to the plane being a dime. I was blown away! I never thought I would get this lucky to have a comparison like that! He wasn't that impressed and even said so on the interview. But I was and he said that, too. I asked him to clarify that he had in fact been shown a UFO by me over the billboard and that it definitely wasn't a star, planet, weather balloon, plane, satellite, etc. He did!

"Finally," I asked one more time, "what did it look like?". He said it was round, shiny silver, and

Heaven Is Space... UP!

spun as it moved.

"Well, wouldn't you say that it matches a flying saucer?" I asked, not able to help myself emphasize again the matching evidence.

He said he didn't know because, personally, he had never seen one. I couldn't argue with that. I was so tired of arguing. Hell, I just wanted the world to be perfect, anyway. What's so wrong with that? God does, too, I thought to myself, but doesn't make it so. I knew better than to touch that—all hell would break loose. It already had.

This is hell; not being able to discuss it, *without all the hell*. Why are religious people so afraid of this evidence? I guess because the same questions really torture them, too. Well, that about rounded out the last of 2006, other than the few sightings I had in Nashville afterwards. It is Christmas Eve now and the Tennessee Titans play today for a chance at a playoff spot. We are all very excited and I still have a song to finish before next week. So, I need to get back on the storyline that left off with

MICHAEL

"CONTACT"!

They have landed! I wrote the ending in my last book. I am not a bible thumper, but I did write that the final battle would be in Megido, Israel. I only did this for reasons of following a fictitious story-line. I do, however, see that the religious community of Jews, Muslims, and Christians revere Israel. The evidence strongly points to them not believing the aliens. The Jehovah's Witnesses are now teaching their flock, like many other Christian sects, that they are demonic. The bible has a killing god who murders innocent children, wants to be worshipped, and plenty of lies. I don't see them changing their "minds." I also don't see an immediate transition from chaos to worldwide cooperation. I think that any if not all of these religious followers will go to Israel. The majority of this population is already there. The ending must entail a slow process of trying to clean-up the earth and re-introduce a whole new way of life into our society, which is now based on money. I have

Heaven Is Space... UP!

followed the evidence for the past fourteen years, since I started down the road of religious examination. And in doing so I have come to one conclusion. The story I heard all my life about the angels being thrown out of Heaven now is very simple to understand. It is a story of lusting for power, to be greater than god. This made me ask the ultimate question; "Who is God?" which is and should be the first lesson in Religion 101. It is, if you aren't already conditioned/"brainwashed" to think God is a given and it doesn't matter what "HE" is. The second should be, "When did it start?" and the third is obvious: "Why would they stay away?" I immediately challenged the magic God because he wasn't here and there wasn't any magic! This was logical, but not for the faithful. In asking these questions I repeatedly found that most of my family and all religious scholars are just that, conditioned to not know or challenge it. Therefore, I set out on my own and ultimately began to study with someone who was also supposedly "looking." However, he

MICHAEL

ultimately couldn't bring himself to prove his magic, spirit, omnipresent god, either! Both the new witness in this story and the old in the first are believers in religion's universal spirit god story. They both still "BELIEVE" this lack of evidence. I can't change them, and my evidence or their lack of it can't, either. They couldn't see the irony of challenging someone to have "PHYSICAL" evidence when they didn't either. Their bible doesn't heal people. I challenged them to do it repeatedly and never saw them or anyone else do it. This led me to a startling conclusion, which came from this simple angel story I learned, like them as a child. Its simplicity stuck out like a sore thumb, that "Heaven mirrored the earth." Wow! I could see our lust for outward beauty and power was ruining our perfect existence as well. Could these angels be gods/aliens who all looked the same and were only capable of power over one another from scientific means like human creations? I was certainly more than convinced of this possibility, when the ancient

Heaven Is Space... UP!

evidence showed statues of god as an alien. His story clearly shows that he knew his "origins" from the universe as well. Adam/Atum is the atom! Yeshua said "we are gods" and was killed for saying he was god. "God" is everything in it, and it came from nothing. John Lennon and Carl Sagan said the same thing. The god story is one of evolution, by which these angels/gods became gods themselves and lusted for outward beauty and the powerful advantage over one another that it would give them. They were able to do this through the creation of man on the eighth day, out of clay, according to the Bible. The sixth day is their evolution from nature/God, who they are. I found that 666 is just the three stages of time, and a metaphor, showing they always make man. The univeral atom symbol has six points. Adams are made up of atoms. He is the biblical quote, "What *isn't*, was, and will be cast into the lake of fire." *Isn't* is the key word. We didn't evolve . . . we're scientifically created. The gods/aliens "evolved"/always existed, just like nature and the atom, "What Is."

All the other creation stories are the same, just like god being from the sky! The Easter Island god is "Make-make," and his creation of man and woman is exactly the same as the Bible's eight-day creation, down to anesthetizing Adam to make Eve. The ancient statues of gods/aliens all look the same and are outwardly ugly. This would make them equal and powerless over one another.

This was all making perfect sense. Outward beauty rules here, and it must be hell. The evidence sure says so. If we mirror them then it must there as well. According to all ancient texts, they also saw themselves evolve and realized that nature created them and the only way to achieve this unique power was through the scientific power of creation itself. They created man for his service and for his ***power through unique beauty of the flesh***. They conquered human creation, space, and "death." They know there isn't an ending to anything, only its image. Our problem and their addiction is just that, an

Heaven Is Space... UP!

image that is destined to doom; science's creation.

Nature creates uniformity within one's own kind. They evolved this way, one species who conquered space. We celebrate and applaud diversity and unique greatness. But it doesn't take a rocket scientist to see our problem, nor this contradiction. The idea of creating bodies for the "soul" purpose of outward immortal beauty is the goal of humanity. If we want to kid ourselves we can, but the religious Armageddon will be all about this one issue. It isn't possible to be intelligent, beautiful, and not know this. Hell, it isn't possible if you're dumb and ugly either. Wanting to be pretty is a terminal disease resulting directly from our mental sexual desire for power. The mother goddess statue makes this so obvious that it is pathetic. The stories of these beings looking "DOWN" on the daughters of man and seeing their outward beauty is painfully clear. They are ugly and wanted to be pretty. This story is the flood story, which is rooted in the universal mother-goddess worship. These ancient statues have heads of aliens and pretty bodies! Aliens are "outwardly" ugly! The evolution of primitive man served only one purpose to them, and that was the vehicle to scientifically upgrade it. Modern man is a product of this scientific creation; primitive man is nature's creation. Mankind is "GOD'S MYSTERY." The evidence says that primitive man's God is an Alien! The trafficking and ongoing scientific creation of humans is why this mystery exists. It can't be stopped. We aren't a "good" thing. Nature is the "good" thing. They don't want to hurt us' they just want to be us! And all for the purpose of sexual power. Most of us, if not all, lust for outward beauty, to have it, be worshipped by it, and last but not least, yes, BE IT! We are sexual "BE"INGS! Actually, I think we all do! I do!! Desire for sex tortures me. It tortures my children who don't want mommy and daddy to be sexual. But the Buddhist monks are the ones who show that they have truly overcome this disease! They are celibate, bald, all look the same, and they don't want to save

Heaven Is Space... UP!

us. Just like the aliens; WOW, these match!

I know I've gotten repetitive, and I'm sorry. But I want to make one last theory about the universe to go along with this ancient "GOD"/ALIEN evidence. On Michio Kaku's homepage, he is accepting people's theory of everything. Mine is solely based on today's knowledge of the atom. The very definition of it is says it can't be created nor destroyed. I have put this on the last page of evidence. Therefore, it has no beginning, only the "THREE" different forms of it. The universe consist of infinite "non-thinking" energy, that is constantly giving birth to matter and repeating it again through its inevitable death. This invisible creator of life is religion's god and irrefutably not a person. Yet, in religion, it is what makes the person. They are one and the same. This is why you have the confusion of god being an individual and everything else, too. It also applies to the multiverse theory, which states that one consists of an infinite "many." And they are infinitely giving birth to "NEW" ones, which is really the recycling of old ones! The structure of the universe is flat, due to the spinning composition of it and the atoms that make it. Even the smallest parts of the atom are held within a flat circular motion due to this centralized giver of life, the nucleus! The membrane theory is supported by many ancient carvings/sculpture of grids with stars and planets. Many even show modern airplanes! This theory does not stand in stark contrast to Michio's umbilical chord theory! As a "matter" of fact, they could be one and the same. The fields of membrane could indeed be spreading out in a round fashion, held together by the gravitational field of its nucleus. It would still be flat. This single field of matter in an infinite universe allows for constant multiplication and the ability to escape the contraction of it. The loophole to conquer the inevitable death of our species would require space travel. They both agree on this "matter," as I do!

That's why it is so scientifically important to see the "UP" evidence of ancient religion. Their

Heaven Is Space... UP!

art might shed some light on the structure of the universe. We must escape the stronghold of matter, by going "UP"! The membrane theory is different from Michio's, in that it doesn't relegate them to be adjoined as bubbles. Michio's does! There is ancient evidence to support this theory. He maintains that the resulting death of a universe will create babies through an umbilical cord. The umbilical cord could be the black hole, which is blowing out matter on the other side. It produces light/"white" matter and is called a white hole. Is it possible to use these to accelerate matter outside the realm of gravitational pull. The yin and yang represent just exactly this premise and that possibility! Like I said before, I had used this symbol to prove the angel/alien story and what they looked like. I now am using it to support Michio's theory. It makes sense to travel through the black hole to find a much younger, "WARMER" universe! It is giving new birth to old matter. I know we can't do this now, but I think it could be possible in the future. Maybe, just maybe, it's the beam me "up" Scotty story! *The Mayan ball game suggests so.* Maybe the aliens are doing it, now! They live "UP" in the sky.

Well, either way, there's no doubt anymore where heaven is. HEAVEN IS SPACE . . . UP! People, for god's sake. We found ancient statues of aliens and flying saucer art. We filmed flying saucers and still can. They are not hurting us. Please give our evidence a chance? "SEEING IS BELIEVING"! IT IS THE PROOF WE ALL WANT.

"FINALLY", you can "SEE" it for yourself in "The Jeff and Mike Show, Real-Life Flying Saucer Hunters!" Send $4.95 (S&H) with proof of book purchase for a free DVD, *Why The Blank Don't They Care*, to: Mike Brumfield, 1066 Golden Herren Rd., Sparta, TN 38583.

Please look for my next book ***The Discovery***. It is not part of this storyline. I will write it as a fiction, based on my dream about the mystery of our species. I've already got the plot "in my head." The inspiration behind it comes from an unknown ad-

Heaven Is Space... UP!

vanced scientific extraterrestrial intelligent being addicted to outward beauty, and my lifelong experience with déjà vu. You might even say it could be my own "guardian angel's" perspective. I'll leave that up to you.

Either way, "again" we all know, that everybody loves a good mystery. Especially me, so please enjoy it. Solving it is my dream and living it becomes mankind's worst nightmare! Little will I know, until the end, that these could be one and the same! Ultimately, this will depend on the way you "look" at it. But remember this: "SEEING IS BELIEVING"! Even if you don't believe it. Follow along as the horror unfolds, if you dare! Don't be afraid; after all, it's only a dream, or is it?

Thank you, thank you so much! Love always, Mike/"Michael"

P.S. Love, forgiveness, and wisdom are one and the same; hate, revenge, and justice are one and the same as well. But they are insane, and as Mr. Spock would say, "not logical."

MICHAEL

Please enjoy the first four books of my story. I started this journey "believing" my parents that heaven was a spirit world. Now the evidence explains to me how their traditions got started and that heaven is "clearly" space/UP today, WAS yesterday, and WILL BE tomorrow.

First book, 2001

Second book, 2003

Third book, 2004

Fourth book, 2005

Reviews

Source Reader (Book 1)
The Two Witnesses and the Religion Cover-Up

"What a fantastic and exceptionally well written book. I feel very confident that my purchase of its sequel as well was a great decision. I can't wait to read it! Thanks for the knowledge Mike."

    Sincerely,
    Robert Salmon
    Ft. Smith, Arkansas

Source Amazon Editorial Reviews (Book 2)
Aliens Gold Tenth Planet

 A shocking sequel to the best seller, *The Two Witnesses and the Religion Cover-up*. Read this riveting story of two men as they continue in their pursuit of answers to the mystery of religion. Then they discover shocking evidence all over the Earth of ancient artwork of aliens and flying saucers. Michael suddenly realizes that religions common themes, mimic today's science and our own pursuits of the same. He theorizes, unlike his religious spirit buddy, that the gods/angels are flesh and blood aliens. They came from up and gave us religious symbols that matched today's science. Their reason for creating us is easy to trace, just follow the gold. The space exploration today could only be made possible through the protection and use of gold. It can also repair and make ozone layers. Hence, our mystery and the answer comes full circle. They came from up and we are going up. All made possible with gold. Creating workers/robots and our recreation of ourselves is more evident today than ever. Up, gold, creating and last but not least our addiction to power through beauty of the flesh, lies behind the shocking discoveries of religion's stronghold. Finally Michael ask the world. Could we be the gods/aliens/angels addicted to the outward beauty of mankind? Head molding, reincarnation and prophecy of our inevitable demise supports this reality. But will we want to be saved from ourselves if we are the aliens? I leave you with a famous quote from the Bible that confirms this possibility, "The angels which kept not their first estate but left it for a strange flesh and keep returning like a dog returning to vomit."

Amazon Editorial Review (Book 3)

Read the riveting ending to the previous two books. *The Two Witnesses and the Religion Cover-up* and its sequel *Aliens Gold Tenth Planet*. What a climax to the most exciting trilogy ever in the history of mankind. Could the aliens be the gods/angels of primitive man? These two desperately try to answer the mystery of mankind. Michael is the only one able to finally break free from his religious upbringing/brainwashing and solve the mystery with religions, common theme of up, gold and creation. Read his spellbinding conclusion to the evidence that clearly shows how heaven became a spirit realm but began in space. Heaven is universally up and primitive man has always looked up to the stars of heaven for guidance from their gods. The quote from Dr. Wendell J. Flanche PhD sums it all up, "The evidence confirms it. Michaels quest to solve mans mystery will astound you. He theorizes that primitive man's ignorance of alien technologies created the magic spirit factor of religion. And ironically it still prevails today, even in the absence of it. This is truly a spectacular scientific discovery!!! Could the most famous man's name confirm his findings that we could be aliens/angels/gods addicted to the outward beauty of mankind? Are we addicted to this power of beauty? I am, Yeshua. Trace mankind's mystery to the pursuit of gold for the gods? Why gold? And why would god need it? What he discovers is shocking because it confirms our past. Gold is crucial for space travel! When we land on another planet they will say we came from up. If we manipulate primates for workers they would say we created them to worship us. If we left them because they became rebellious with their new-found knowledge they would say we were magical and not understand our need for gold. Ultimately it would take thousands of years for them to solve this mystery, but in the meantime they would still have an economy based in gold. This is our history! They "MATCH"! What an unbelievable achievement by this author. It appears he has solved the ultimate mystery; mankind, religion, and the universe itself!

Source: Author (Book 4)

The evidence of our ancient past reveals primitive man's universal religious stories of people in the sky that created Him. His creation was meant to serve their needs. We now live in the sky and use gold to do so. These match! He made statues of these people and they are clearly the stereotypical alien in His fiery chariot, which resembles a flying saucer. Unbelievable? Believe the evidence! It never lies. The Dresden Codex of the Mayan obelisk reveals that contact in 2012 is brought on by the ravages of global flooding. Two words . . . global warming!

Global comprehensive evidence that religion originated from flying saucers and their alien occupant.

THE VIDEO/PICTURE EVIDENCE

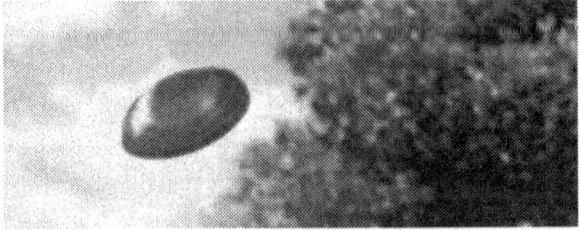

Paul Villa photo of flying saucer circa 1960.

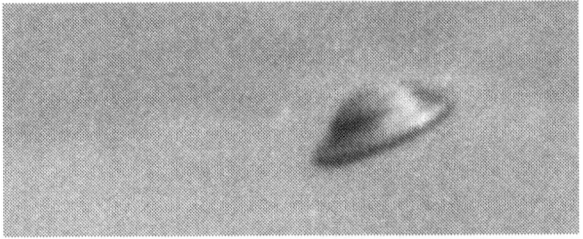

Picture on front cover of third and fourth book. How can these match when they are taken twenty years apart. Again, these "MATCH" ancient cave drawing on front cover. World's largest saucer on head of Easter Island, and aborigine saucer and alien's gold halo above (back cover) protecting it in space.

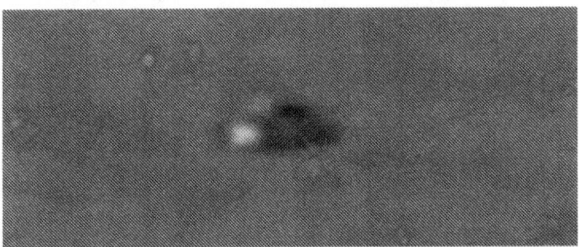

Video taken in 2003 by Jeff Willes of Phoenix, Arizona. To buy video, type his name in computer or call 623/847-9132.

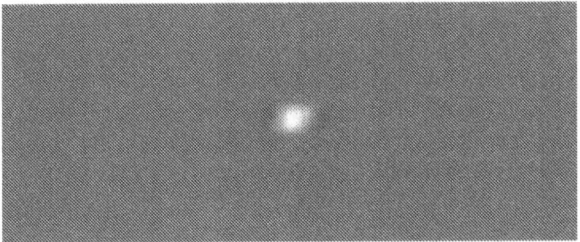

Video taken in 2006 by Mike Brumfield in Phoenix, Arizona.

Ancient Evidence

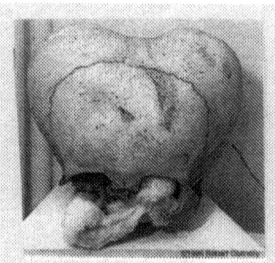

Oldest rock art on record catalogued by the Leakeys, ca. 50,000 years old, from Africa. Clearly shows little alien highlighted in box. It also shows another taller species restraining one of their own. The two heads above it were found in modern-day Israel, and are Sumerian; approximately 10,000 years old. The skull (from Peru) also supports ancient rock art of aliens in Africa. There are universal religious stories of two creations of man. Does this give us proof that they first tried to manipulate their own species to serve their needs?

Oldest Sumerian/Ubaid "God" statues on record in Museum of Antiquity, Cairo, Egypt. Clearly shows male and female gender and alien-looking beings. Picture, lower left, even shows mother nursing baby. Zechariah Sitchin claims these are android robots. They are, for God's sake, real "PARENTS!" If these are the most ancient statues that don't look like us, could they be primitive man's universal god/angel? Look at "Mother Goddess" statues on the following pages. They clearly are the god/angel that mixed with the "pretty daughter" of man.

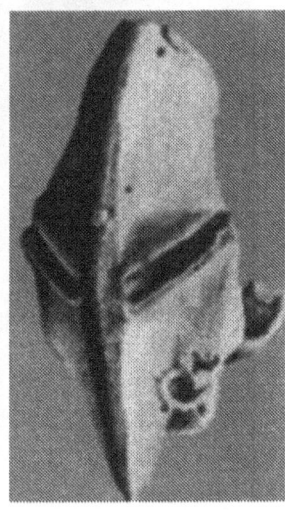

Ancient Sumerian King and Queen and Priest. Notice Priest has bald head—indicative of Alien God. Also notice big eyes.

The Sumerians carved statues of the gods from stone. From the statues we can see what they thought the gods looked like. Many gods looked like short people with round bellies. They had thin lips and big noses. They wore skirts made of sheep's wool. In fact, many statues of the gods looked like the statues the Sumerians made of themselves!

What Did the Sumerians Believe About the Meaning of Their Lives?

The gods of Sumer looked like men — and they acted like men. The gods liked good food and nice clothing. They got married and had children. Sometimes they were kind. Sometimes they were cruel. Either way, the Sumerians believed they had no control over what the gods did. Rather, the Sumerians believed that they were slaves of the gods. This story tells why.

THE SUMERIAN STORY OF THE CREATION OF MAN

The gods had always worked for a living. But when the goddesses were created, the gods had to work even harder to keep them happy. Then the gods had great trouble getting enough bread to eat

This is the oldest known historical record of Mankind's "God" story. It clearly shows they were real flesh and blood people. It also shows they created Mankind to work for them. This is an excerpt from an educational text called *Ancient Civilizations*.

Mother goddess statue from Catayal (modern-day Turkey), ca. 8000 B.C. It clearly shows an alien head representing God and a beautiful woman. This story is also universal in religion and reflects Gods/angels mixing with "pretty" daughters of man.º

These were found in Iraq. They are ancient statues of gods. Proves they have children, just like statues of alien parents holding child.

Cloning! Two squiggly lines, DNA!

These look like head of mother goddess statue and again connects aliens to Pyramids like evidence from Erich Von Daniken's book, *Gold of the Gods*.

More examples of "God" statues emphasizing "Big Head" and eyes. See how indentation in the forehead matches Iraq's statues. The top one (from Iraq) even has six fingers like other succeeding "God" statues. The bottom one is from South America.

These gold artifacts are from Erich Von Daniken's book *Gold of the Gods*.

Gold artifact from ancient mine in Peru. Notice two alien-looking beings holding snakes and third one at top inside pyramid. The circles look like flying saucer stem cells, atoms or eggs that are fertilized.

This connects pyramids to aliens. The pyramid served as image of rock that contained gold, quartz.

Alien-looking figure, right, has pyramid on head, penis and snake/DNA halo. The Bible says (Luke 6:4) "Only the Father in Heaven is God." Are these the fathers of heaven? Erich theorizes the skeleton on left could be a coded disk for message to contact future man. It is made of aluminum and coated in gold. We sent a coded disk into space with the same composition. I, again, see DNA, cells/eggs and chromosomes. The skeleton's head has a halo around it. See horned face like Africa rock art and Israel statures. The skeleton represents our deadly creation.

This is the most important evidence because it shows us, what their gods looked like (aliens), what they came and live in (flying saucers) and why they needed gold (space life and exploration). Hence we have the Biblical quote "Heaven's streets are paved with gold." It is a universal religious theme. Heaven is space, up to primitive man.

Cave drawing of aboriginal god Wandjina. Notice similarity to Owl Man. Also see halo around above head. This supports 10th planet story of gold replacing ozone and "who" was mining it before they created man as a "tiller of the ground" in Genesis. Man's purpose supports skeletal discoveries in gold mines. Gold protects astronauts from dangerous life radiation in space.

MOST IMPORTANTLY This solves pyramid mystery. Gold is found most in quartz which forms natural pyramid shape.

Religious statues by Olmecs from Mexico. They are also known for mysterious carvings of huge heads!

Figure circled is made of red lava rock like Easter Island man doing mystery. This is what power struggle of gods is about, us. Red symbolizes creation. Notice opposing sides black and white like Easter Island Man and Yin and Yang. Notice six obelisks like stones. Coincidence? Don't think so. Notice similarities to Easter Island statues, Israel statues, aborigines, all other "big headed" God statues from every continent.

The figures are all black and white "facing" each other. The one in the back that is porous looking is the only red one. Could this represent the power struggle over mankind's inevitable creation and does it involve the sixth chromosome? Also, notice the clear resemblance to the alien statue from Israel and the head on the mother goddess statue as well. It clearly looks like the Easter Island heads except for the elongation. They are the product of the mix between the sons of gods/aliens and man. I think losing the bulbous head was the first indicator of their pursuit toward outward beauty. The story supports this with their reason for mixing in the first place. (See back cover)

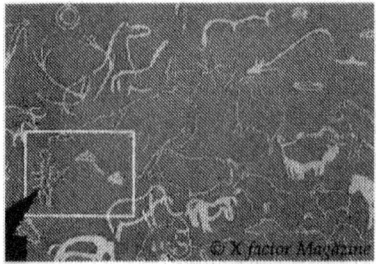 

Japanese discovery on the island now known as Taiwan. It is a drawing by a general that discovered a strange ship on the island. This happened in 1806! Notice the ships drawn above it. He found this drawing on the hull of the ship. The writing looks like the hieroglyphics found on the aboriginal gods halo and supports the description of Roswell's.

The cave art clearly shows matching saucers. This is dated circa 15000 years old. It is in France. It shows smaller ships coming out of a large one and abduction! The lines represent the invisible energy taking the human up. The top right picture is the oldest rock of aliens from Africa date circa 50000 years old. It shows the little alien/Roswell gray in control observing. He's even protected by a box that looks much like a tree trimmer's carriage. The others are larger and restraining one of their own. They must have first made themselves larger to be more able to control their scientific manipulations of primitive man. They are obviously serving the little guy and they are struggling with one of their own. Anyway, this supports the scientific manipulation of themselves. See the one with the horns. Is this what gave us the first images of the biblical "devil". Read on! See similarity to statues on following page.

These reptilian looking skulls are found in Ubaid, Iraq. They look like the reptilian looking tall ones on the rock art of the previous page. Scientists have repeatedly mistaken the eyes for sunglasses or goggles. However they are very similar to the large slanted eyes of the Roswell gray alien. They are identical to the eyes of the mother goddess statue from Israel. Notice one is a divided looking skull, giving it the appearance of hornlike appendages, while the other is elongated. They are clearly two different types. Were these a product of the first attempts to make themselves larger for power or for mining gold. Anyway, it is clear here and in the writings that scientific creation was producing things like this, the mothman, centaur and other abnormalities. The little guy with a HUGE head is from Utah! See appendages (Devil's horns?).

ALIENS AND UFOs

The Starchild controversy

SINCE FEBRUARY 1999 a bizarre looking skull, known as the Starchild skull, has been exhibited at UFO conferences and heavily discussed in UFO journals.

The Starchild skull is alleged to be the remains of an alien-human hybrid.

Legend of the Star People

According to the Starchild Project, an organization that wants to arrange DNA testing of the skull to prove an incredible origin, the skull was discovered in the mountains of northern Mexico. Indian tribes from the region have legends of Star People – beings from the sky who visit Earth to impregnate local women before returning years later to retrieve the hybrid infants.

DEFINITION

A hybrid is a cross between two different breeds or species. Only closely related species can interbreed or "hybridize," and it seems unlikely that humans and aliens would be similar enough.

Big head

The skull has several strange features that suggest it is not human. It has a massive brain capacity, flattened rear, shallow eye sockets, and is missing the front sinuses.

The Starchild Project claims to have consulted over 50 experts, the vast majority of whom argue that the skull is that of a deformed human child.

Most experts say that the Starchild skull is that of a child suffering from hydrocephaly, a disease in which fluid builds up on the brain and makes the skull swell.

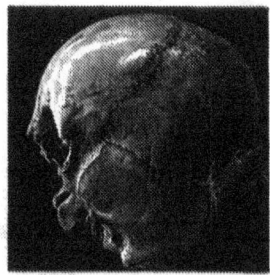

■ *The Starchild skull is far from normal. But is it from an abnormal or cradle-boarded human infant, or perhaps an alien-human hybrid?*

It is also widely thought that the skull has been cradle-boarded. Cradle-boarding is the practice of strapping an infant's head to a board and causes flattening of the back of the skull. It was practiced in the area of Mexico where the skull comes from. The Starchild Project argues that close examination of the skull rules out this explanation, and is attempting to raise funds to pay for DNA testing – the only way to be certain of the skull's origins.

"Beings from the sky" Is this the Owl Man which looks like an alien? He is pointing up! The skull supports this reality. Also "impregnates women" supports the cover art of the mother goddess statue, yet also has a big head! Skull is evidence of aliens being flesh and blood and these "sons of God" in Genesis 6:4. It supports my theory that they are not religious "spirit" magical beings. However, I conclude the atom, which makes everything, is "religion's invisible spirit" creator, evolving scientifically through time, not by magic. Read on. It scientifically fits religion's invisible omnipresent God.

Flying saucer

Giants of Easter Island South Pacific

Notice the saucer on top of head tells us where they live just like owl man, aboriginal god, and Starchild legend spaceships just like on covers! They live up in saucers! Six strands of rock looks like DNA readouts. It also could implicate the sixth chromosome mystery or the Jewish creation on the sixth day. Giants were the offspring of gods and "pretty" daughters of man. This is when our separation occurred because wickedness spread all over the earth. A great flood followed. This is a red figure that symbolized mankind. See how he is doing the mystery or transcendental meditation, and looking up!

The six strands of rock below the Alien looking god could symbolize a DNA readout. I am intrigued by it being six strands. The day of man's biblical creation is the 6th. The Hopi prophecy has six beings (five of man, one of an alien). The hummingbird of the owl man in Nazca reflects this theme as does the biblical "devil". Is it possible that the sixth chromosome is the source of this "looks" manipulation. I've been reading a fascinating book called "The Sixth Chromosome". There are many other things pointing to creation involving the number six like the atom and Jewish star's number of points. The planet mars is the sixth from the tenth. There's more read on! Also, look at man meditating/doing mystery. He is made of red lava rock and is similar to many other representations of first religious worship. Red also represents blood and creation. Mystery worship is universal from Buddhism to sitting Indian style. Also notice black moai that looks somewhat different from the whiter standing ones. This parallels black and white yin and yang. It also parallels Olmec statues.

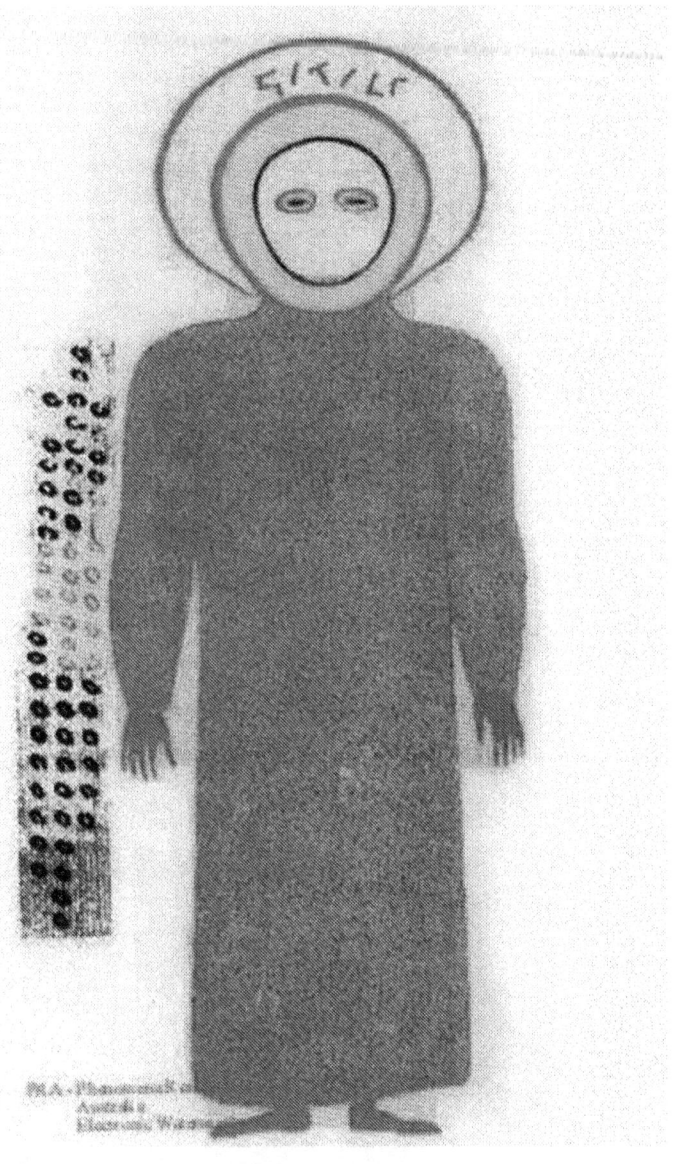

Aborigine cave drawing dated circa 30000 years old. Notice hieroglyphics on gold halo. Also notice readout similar to Easter Island one. These rock layouts/DNA readouts are common across the earth. Looks like scientist/astronaut in robe!

The bronze statue is from Kiev and is circa 8000 years old. It has six fingers supporting the existence and authenticity of the Roswell Alien autopsy. The recovered dead alien had six fingers and toes. This figure also supports the need for gold as protection in space. It has a Halo. Compare it to the following aboriginal gods. They look alien and even have gold painted halos around their heads. The Aztec block shows two hands intertwined with six fingers. These hands alone represent their god's creation of them and the entanglement represents DNA, how they were created. These match our medical symbol, intertwined serpents. How can they match when it takes an electron microscope to see them? The gods must be scientifically advanced!

Australian rock art compared to same in Utah.

In the Beginning

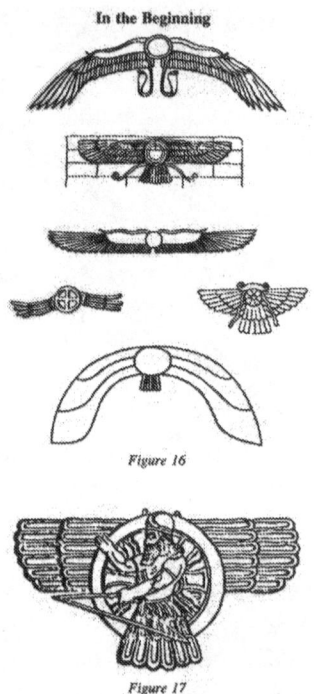

Figure 16

Figure 17

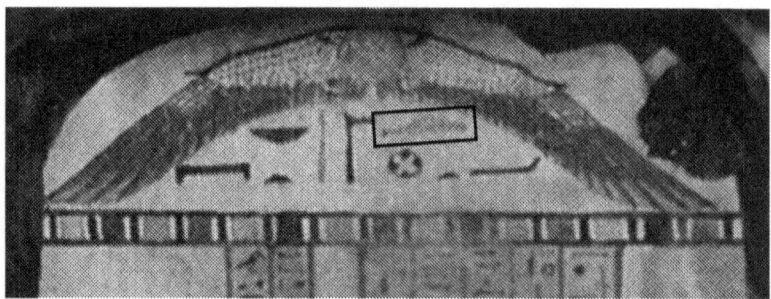

Ancient God symbolizes a flying saucer! These are all ancient symbols for god from Sumeria, Babylonia, Assyria, to the Egyptian one at the bottom. It has an actual saucer below the omni-present symbol of god which could literally be called a flying saucer. They all share this characteristic! Now we know why they're everywhere. But remember the atom is also a circle that is "every-thing". The Egyptian god is Atum! Notice the snakes for DNA creation. Also notice the cross symbol. It is the oldest geometry on earth representing the 10th planet. The Assyrian one clearly shows how man put himself in the circle. He became god!

Famous NASA Tether Incident

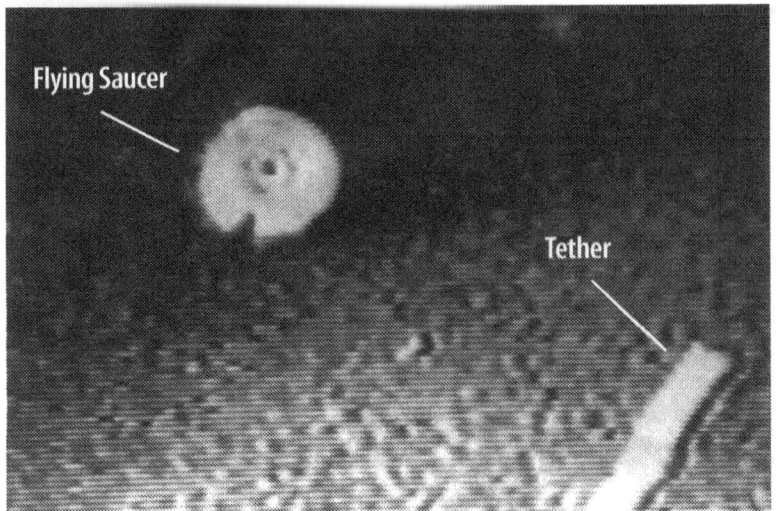

Saucer looks like galaxy, yin and yang, and the atom.

Galaxy has black hole in center which emits white matter. See yinyang similarity.

white hole

black hole

Cave drawing showing saucer moving up

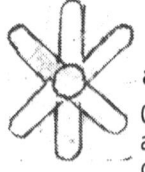

atom

On a flat surface all would show center protrusion.

Could spinning be key to anti-gravity? Is Event Horizon proof time can be stopped? Can all life be related to atomic structure? Does knowledge of atom answer life's mystery?

Famous NASA Tether Incident

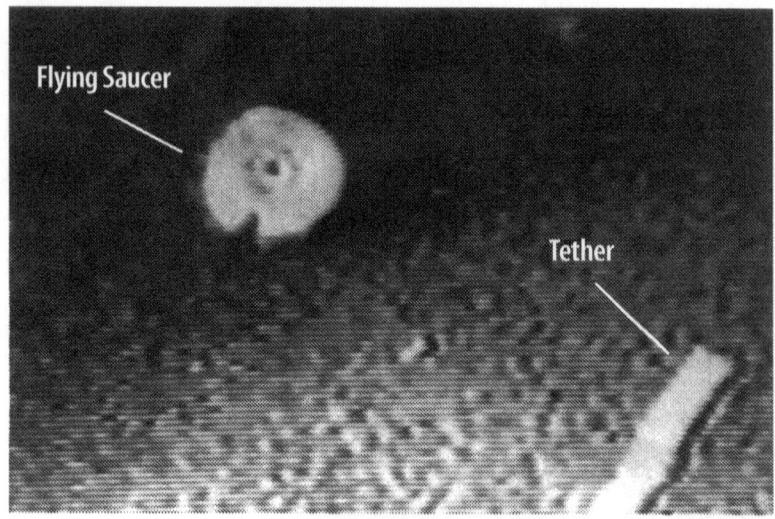

"Ancient" bronze disk from Norway. Gold overlay of sun, moon and objects in sky. The holes on outer perimeter look just like ones on disk from Turkey, Ohio and notches from Dropa stones. Notice seven circles like atoms between sun and moon!

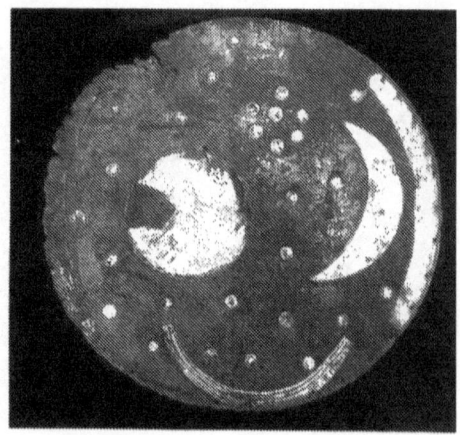

Notice the similarity of disk sculpture with indentation to flying saucer photograph by NASA. Most importantly, sculpture is in gold, which is what we use in the construction of our spacecraft today.

"Ancient" bronze disk from Norway. Gold overlay of sun, moon and objects in sky. The holes on outer perimeter look just like ones on disk from Turkey, Ohio and notches from Dropa stones. Notice seven circles like atoms between sun and moon!

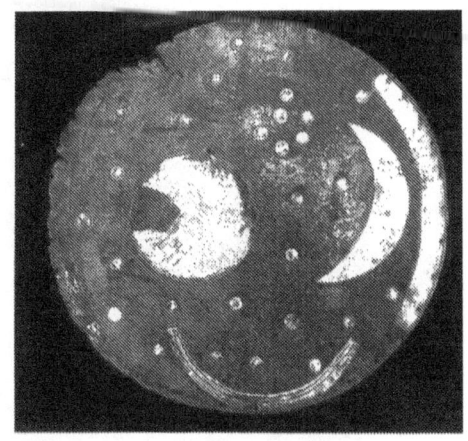

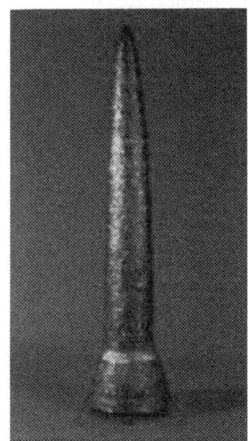

Mysterious ancient gold cone hats of Europe. They look like rockets and have flying saucers images as well as moons and suns. The priests wore these as hats.

This ancient Egyptian sculpture of Ahknenton has a religious ceremonial hat that resembles the gold cone and Easter Island.

Again, these disks, hats, gold, and obelisks all implicate and match flying saucer evidence and explain why they depict stars in space. It is already conquered by space faring beings that created our mysterious species to mine gold. Gold is crucial to explore space!

From 6000 B.C. A plate from Nepal, the decoration shows a saucer-like shape and a large-headed humanoid. These are craft seen by many thousands of people today all around the world.

UFO Coin, 1680. French Medal apparently commemorating a UFO sighting of a wheel-like object in Renaissance France.

The Yappese are the greatest of the Polynesian navigators. Our Hawaiian voyaging canoe "Hokulea" has a Yappese navigator. Since Yap is geologically unique in Micronesia, sedimentary in origin, all the rock is shale. Palau, about 700 miles southwest of Yap is predominately volcanically uplifted limestone created from ancient coral reefs. It is uniquely crystalline in nature. Voyaging to Palau by canoe, Yappese quarried this stone, risking their lives to get home with the largest coin. Many voyages were fraught with danger and adventure. The tougher the voyage, the more the money was worth! See hole in the center, like other ancient saucer/disk statues.

In Jabbaren, in the Tassali mountains, Algeria, south of the Hoggar. A 6 meter-high character with a large round decorated head. The massive body, the strange dressing, the folds around the neck and on the chest suggest some ancient time astronaut. A similar character is painted at Star in the Tassali, in the Cabro caves in France and in several other places. Some of them are much smaller and raise their hands towards a giant being, of non human appearance, sometimes these "round heads" beings seem to hover in the air. On right, an ancient painting ca. 1700 A.D. See how all the other ancient saucer art and photographs have a hole in the center. The biblical description of these fiery chariots, a circle within a circle!

Ufologist Bob Dean noticed a similarity between this UFO photographed by police officer Mark Coltrane in Colfax, Wisconsin on the 19th of April 1978 and the object in the ancient De Gelder painting.

Fig 7: The Baptism of Christ by Aert De Gelder was painted in 1710 and hangs in the Fitzwilliam Museum, Cambridge.

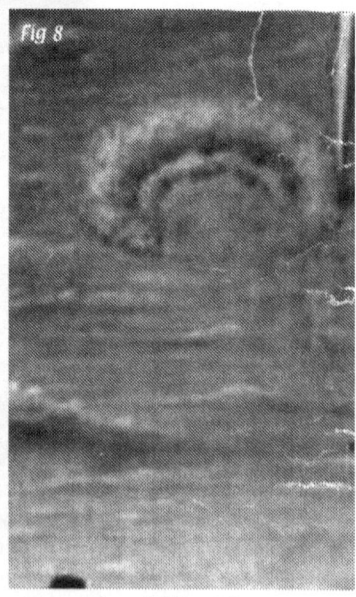

See similarity of photograph by Ed Walters on left to ancient painting of UFO on right.

Religious ancient disk and obelisk on left from China. See three dots at top of obelisk. Does this represent atomic propulsion of today's rockets? Ancient clay disk on right from Turkey.

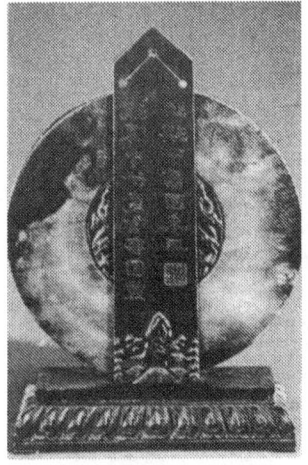

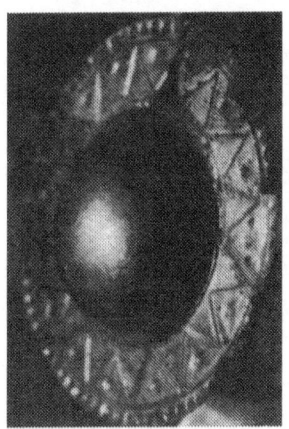

Ancient gold disk from Bogota, Columbia, matches clay mold from Turkey and flying saucers' shape.

See notches on outside perimeter! Important because it matches many others from other countries. All ancient.

Gold disk from Peru gold mine. See sperm and alien head and face. The center is full of atoms . . . again possibly evidence of atomic propulsion. There are faces in sun and stars. We are stardust/atoms/adams. Notice diamond infinity symbol.

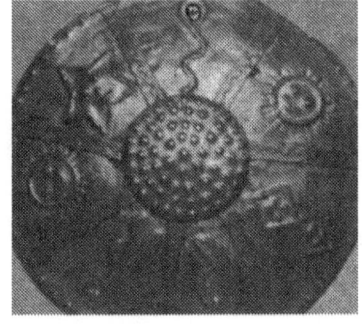

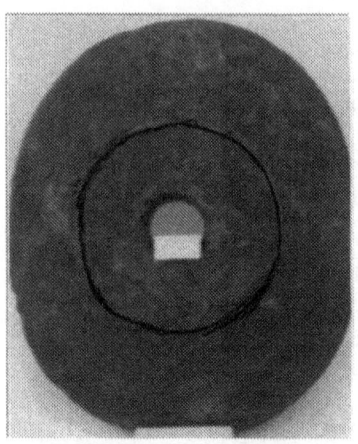

Ancient Yap stones "money stones" from the South Pacific. These are still used for money today, although it is rare. They look like the NASA tether saucer and all other ancient disks.

Ancient Chinese gold disk with rubies. It says "anywhere the sun shines, life will exist." This looks like a computer chip. The disk fits disk on Genesis probe.

Ancient cave art from Ubekestan. See notches like in Ohio and spirals like dropa stones. Besides, he looks like an astronaut. See atom symbol on jaw line.

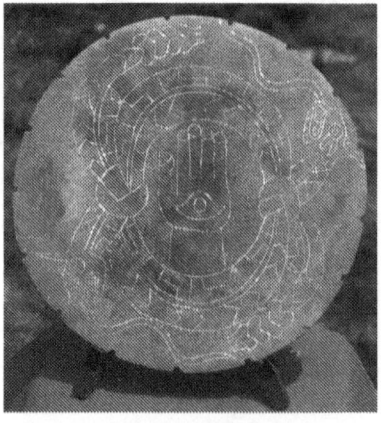

This ancient disk was found in Ohio, my home state. It is a beautiful piece of evidence to support my theory. It has an eye in the middle of the "right" hand which is center of disk. This clearly represents the gods being in control, knowing all (omniscience) and controlling all (omnipotent). The rattlesnakes represent our deadly creation. The disks give them omnipresence.

Looks just like other UFOs of NASA!

These are dropa stones from Tibet and are circa 10000 years old. They were found deep in a cave with the remains of about 400 skeletons of little people with big heads. The island of Yap values an identical stone as money. They are called money stones. The largest ones measure up to 10 feet, and are made of polished white limestone. The whiter they are the more valuable. Now we see where the white thing comes from in religion. If these gods stay in spaceships they would be really white looking. The universal alien is the Roswell gray! If you don't buy this then buy an alien doll. It will be him!

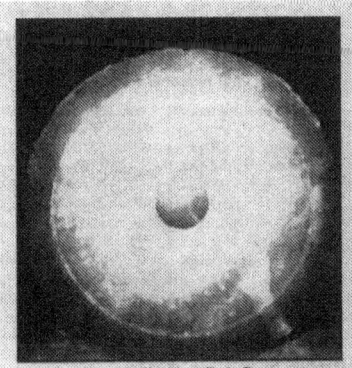

Dropa stone: Artifacts called Dropa stones, which bear an uncanny resemblance to the UFOs involved in the tether incident.

Last but not least, the Legend tells how they were attacked and eventually killed off by neighboring tribes because they were so "UGLY." Here again is evidence why they can't cohabit with us and how language comes full circle to support the evidence and answer the big question, "Where did they come from; What do they look like and why do they stay away?" They "dropped" out of sky according to legend and this is why the tribes they spawned are called Dropas. They still exist today and have physical attributes that resemble the alien. The Owl Man is your next answer and Cernes Giant the last of the three WWWs. Where, what, why!

Ancient stone carvings from Peru, "Ica stones."

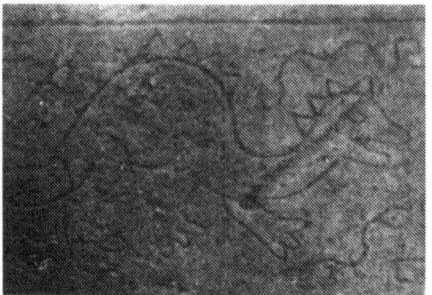

Notice sperm-like objects with DINOSAUR! Also amulet on right clearly shows big headed alien god above earth and not on it. The earth is gridded like we do today with latitude and longitude lines. How is any of this possible without space already being conquered by scientific beings?

Giant stones of Costa Rica. Again why? Does this prove knowledge of planets and atoms that would be important to space-traveling gods?

MONUMENT 4,
LA VENTA
2.26 m (7.41 ft) tall.
La Venta Park-Museum,
Villahermosa.

HEAD 10,
SAN LORENZO
1.8 m (5.9 ft) tall.
Tenochtitlán Community
Museum, Veracruz.

These are six-foot giant heads of the Olmecs. Giant Heads! Notice DNA symbol and cross symbol of tenth planet on jaguar head at left. This mirrors sphinx. The right head has "six" claws on forehead and symbol of atom, cell or fertilized egg and flying saucers!

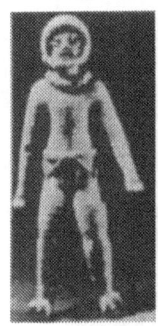  Maya god has gold halo above his obvious spacesuit. This makes it clear, gold is used to protect us in space! This evidence speaks for itself. These are ancient astronaut statues compared to a real one. The bottom one is identical. It was found in Peru and is 6,000 years old.

Also, these are airplane statues made of solid gold. This supports space travel's need for gold and proves gods are flesh and blood beings who have already conquered space!

THESE SPEAK FOR THEMSELVES AS WELL. They are all made of gold. From Egypt to Peru!

India

Look at alien eyes!

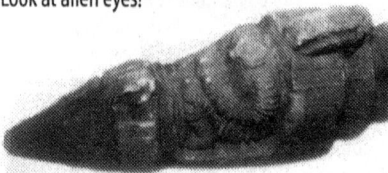

Turkey

Look at matching thrusters on obvious rocket. The head is missing on pilot.

More ancient statues

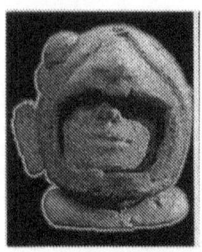

Ancient European astronauts!

Mexico Gold Star God
Alien head on DNA from sun bottom left hand.

Brazil
Kayapo tribe still celebrates the legend of "Teacher from Heaven" Bep Kororot; this is his suit; stick, his "fire" stick. He made the villagers' weapons turn to dust when they tried to "attack" him. He helped them and then went to mountain top and disappeared in a cloud of thunder. They await his "Return!"

There are many legends of gods from the sky, in clouds that make thunder. The Kayopo story on the previous page mirrors that of Moses and his "ten commandments." They were also to teach us. Could this be the cloud of thunder? Actual photo taken by Army private in 1965. Eye witnessed by others and never explained!

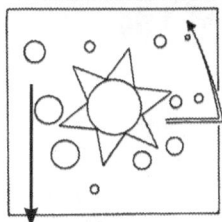

Flying Saucer

Cave Astronaut

10th Planet

1. Notice cave grid looks like computer grid of space, and is flat like floating membrane. The membrane theory says space is infinite. It is a cyclical dance of creation and destruction. Also star of cave sculpture just like the Sumerian clay tablet below, and atom symbol. The clay tablet is dated circa 13000 years. The cave drawing is much older. They both show a 10th planet in our solar system. HOW?

2. The cave astronaut and gemini-looking capsule are also ancient. This is proof they existed before and supports my conclusion. Read on!

3. This is "Matching" irrefutable evidence that the ancients were communicating with space-faring people! Their absence makes it clear. Our mystery is what they look like!

4. See flying saucer in sky above astronaut! Again, a hole in the center!

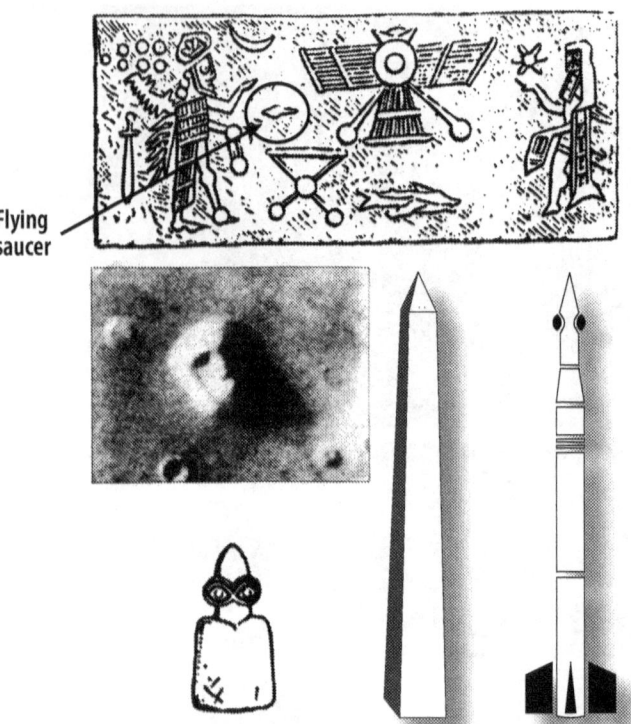

Flying saucer

1. Notice the satellite on the clay tablet going from Earth (7th planet) to Mars (6th planet). It looks just like ones today. This tablet is also circa 13000 years. Also notice symbol for Mars matches atom and Jewish star. Is this proof that Mars could have had Man there first and we destroyed it with nuclear weapons? The "man" on mars is in a suit. Is it reason for contact?

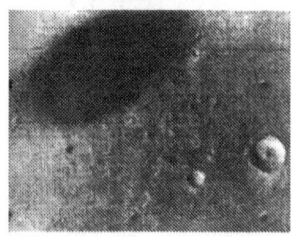

2. See how the helmet of Mars "man" on tablet matches our pictures of face on Mars.

3. Notice ancient satellite looks like alien head and eyes of nuclear missile. Egyptian obelisk matches nuclear missile. Egyptians called obelisks "rocketship".

4. See flying saucer monitoring earth on mars clay tablet.

5. This is a photo from phobos satellite sent to view mars moon phobos. It is irregular shaped and appears to be hollow. Could it be used as a space base on the inside? We think asteroids could be used this way as natural spaceships. This is the last picture it took before it was deemed "destroyed" by space debris. Looks like a flying saucer to me.

Three famous UFO incidents in the U.S. reported on the front page of each respective city's newspaper. The dates and places are on the last two. The first is Los Angeles and shows us shooting at it. It happened in Feb. 25, 1942. No wonder they don't cohabit with us. We never recovered it. Ten innocent civilians died from the shrapnel fallout!

286   GENESIS REVISITED

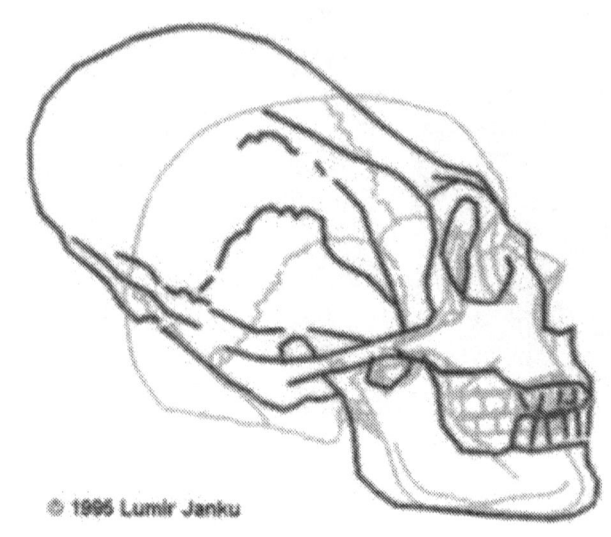

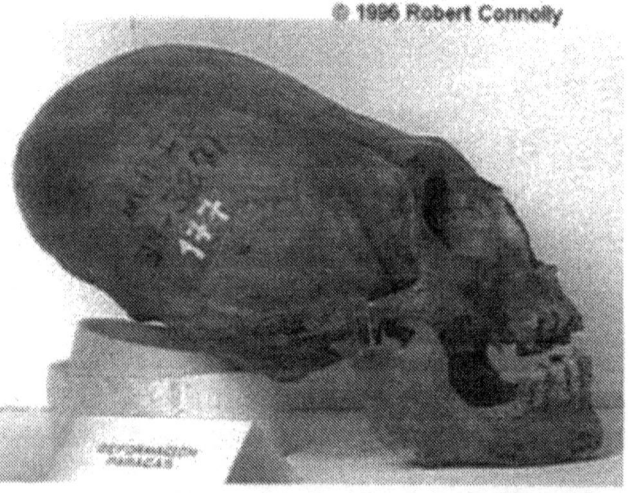

Head molding was an ancient universal religious practice! Obviously, they were trying to imitate their gods' appearance.

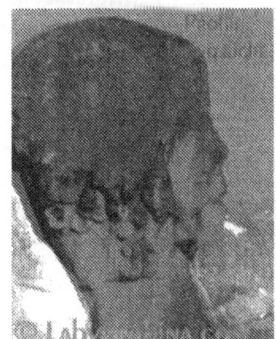

Ancient sculpture shows alien head and man's head together!

Ancient religious sculptures that match from three different continents!

Ancient Moai Carvings
Japan — Marcahuasi Peru — Easter Island

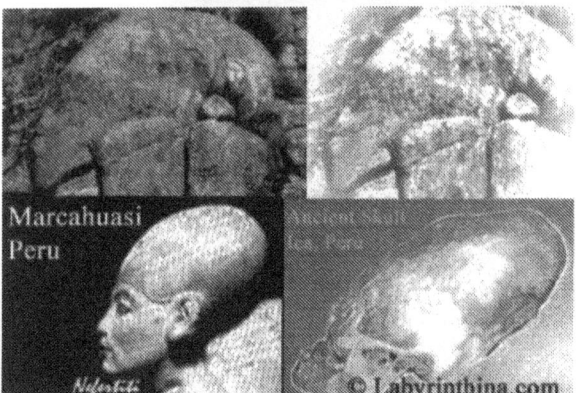

More matches! Ica, Peru has Ica stones that tell and show my theory. Alien-looking gods here with primitive man and dinosaurs creating our species, the mystery/modern man.

Ancient Egyptian relief "Stellae" of Ahknenton, Nefertiti, and children. Notice bald elongated heads and big alien eyes!

Egyptian papyrus clearly show Ahknenton's head without hat. It is alien looking like long limbs and fingers. The bald head is universal religious practice like ancient head molding!

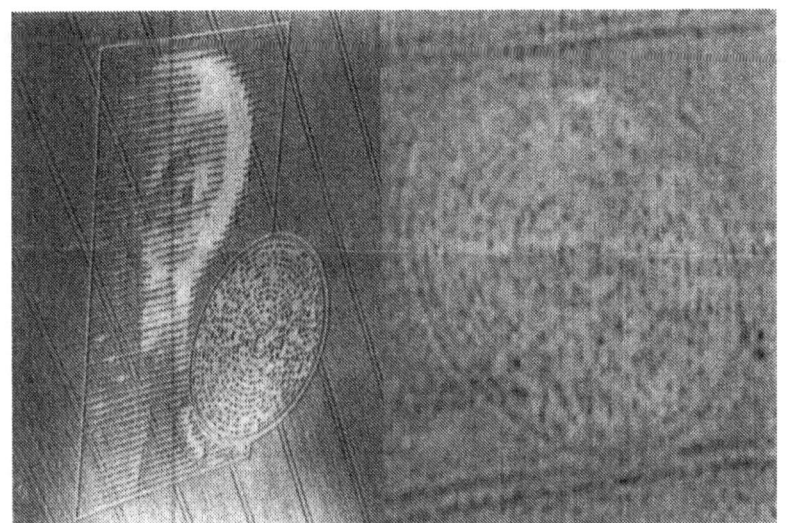

This crop circle has a coded message: "We are the good guys, not mankind!" This matches the quote by Yeshua when they called him good. "Only the Father in heaven is good," Luke 6:4. Is this proof that the aliens are the "father" and it is a plural term also. Yeshua said, "The Father and I are one." He also said we could be too! The evidence of an alien with a penis from Erich Von Daniken and the head of the mother goddess statue proves this universal story's facts.

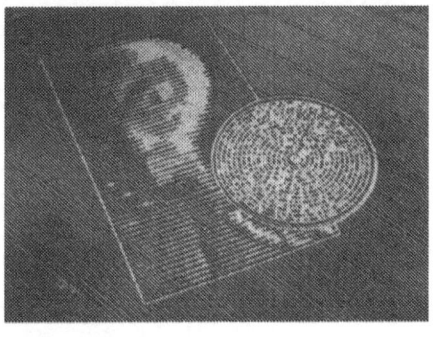

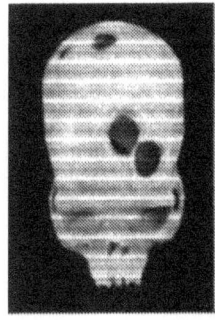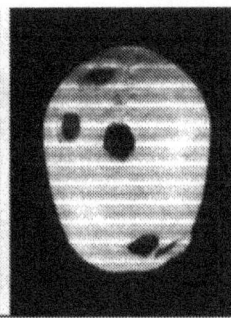

Ancient intrepanation skull! This was also universal religious practice. This could be how mental telepathy works. We are doing this today with cybernetics. Surgical tools found with skulls were made of gold!

Gold is the most ancient metal known to man and sacred to religion, everywhere. This information is provided by World's Leading Gold Mining Co.

History of Gold - Timeline

4000 BC	Gold is first known to be used in parts of Central and Eastern Europe.
3000 BC	The Egyptians master the arts of beating gold into leaf and alloying gold with other metals to variations in hardness and color. They also develop the ability to cast gold, using the lost-wax still used in today's jewelry industry. The Sumer civilization of southern Iraq uses gold to create a wide range of jewelry, often us sophisticated and varied styles still worn today.
2500 BC	Gold jewelry is buried in the Tomb of Djer, the king of the first Egyptian dynasty, at Abydos.
1500 BC	The immense, gold-bearing regions of Nubia make Egypt a wealthy nation, as gold become recognized standard medium of exchange for international trade. The Shekel, a coin originally weighing 11.3 grams of gold, is used as a standard unit of mea throughout the Middle East. The coin contained a naturally occurring alloy called electrum, \ approximately two-thirds gold and one-third silver.
1352 BC	The young Egyptian King Tutankhamen is interred in a pyramid tomb laden with gold, his re an extravagant gold anthropoid sarcophagus.
1350 BC	The Babylonians begin to use fire assay to test the purity of gold.
1091 BC	Squares of gold are legalized in China as a form of money.
560 BC	The first coins made purely from gold are minted in Lydia, a kingdom of Asia Minor.
58 BC	Julius Caesar seizes enough gold in Gaul (France) to repay Rome's debts.
50 BC	The Romans issue a gold coin called the Aureus.
600-699 AD	The Byzantine Empire resumes gold mining in central Europe and France, an area undevel fall of the Roman Empire. Artisans of the period produce intricate gold artifacts and icons.
1100	1100 Venice secures its position as the world's leading gold bullion market due to its locatio trade routes to the east.
1284	Venice introduces the gold Ducat, which soon becomes the most popular coin in the world, so for more than five centuries. Great Britain issues its first major gold coin, the Florin, which is followed by the Noble, the A Crown, and the Guinea.
1511	King Ferdinand of Spain sends explorers to the Western Hemisphere with the command to '
1717	Isaac Newton, Master of the London Mint, sets price of gold that lasts for 200 years.
1787	First US gold coin is struck by Ephraim Brasher, a goldsmith.
1792	The Coinage Act places the young United States on a bimetallic silver/gold standard, definin Dollar as equivalent to 24.75 grains of fine gold, and 371.25 grains of fine silver.
1803	North Carolina site of first US gold rush. The state supplies all the domestic gold coined for the US Mint in Philadelphia until 1828.
1848	The California gold rush begins when James Marshall finds specks of gold in the water at J sawmill near the junction of the American and Sacramento Rivers.
1850	Edward Hammond Hargraves, returning from California, predicts he will find gold in Australi week. He discovers gold in New South Wales within one week of landing.
1859	The Comstock Lode of gold and silver is discovered in Nevada. As a result, Nevada is mad years later.

Oldest gold mines found in Africa civilization traced from NE Africa science traces our origin to hominid they named "Eve."

Genesis Project

Capsule bearing solar secrets

By PAUL FOY
Associated Press

SALT LAKE CITY — In a harrowing feat high over the Utah desert Wednesday, two helicopter stunt pilots will try to snatch a floating space capsule that holds "a piece of the sun" and bring it safely down.

Their biggest fear: What if they flub it on live TV?

And that's entirely possible. The pilots rate it 8 or 9 on a difficulty scale of 10.

"It's like flying in formation with a giant floating jellyfish," says pilot Dan Rudert.

The stuntmen will be trying to hook the 400-pound Genesis capsule as it hurtles 400 feet a minute. Inside it are fragile solar wind particles — so small they're invisible — which scientists hope will reveal clues about the origin of our solar system.

The biggest challenge, pilots say, will be flying at 40 mph almost a mile above the desert without visual reference points to judge distance or speed as they close in with hook and cable.

The helicopter pilots will have five chances to snag the capsule in midair. Military pilots were unavailable for a mission that required them to commit to a task six years in the future. The civilian pilots have replicated the retrieval without fumbles in dozens of practice runs, but are terrified of failing as NASA television broadcasts a worldwide feed.

If they miss and the Genesis capsule hits the ground hard, scientists say they'd have to spend months sorting through broken jewelry-studded disks holding the tiny solar wind particles.

There are other opportunities for the $260 million mission to go awry, too. For NASA engineers a white-knuckle moment will be when the capsule must be steered through a "keyhole" high in the Earth's atmosphere. If the experts at California's Jet Propulsion Laboratory can't line up the precise entry and angle, Genesis will be waved off on an elliptical orbit of Earth, and another attempt would be made in six months.

The Genesis mission marks the first time NASA has collected and returned any objects from farther than the moon, said Roy Haggard, Genesis' flight operations chief and CEO of Vertigo Inc., which designed the capture system.

Together, the charged atoms captured on the capsule's disks of gold, sapphire, diamond and silicone are no bigger than a few grains of salt, but scientists say that's enough to reconstruct the chemical origin of the sun and its family of planets.

Scientists will keep busy for five years after Genesis completes its ride in. It will take at least six

> Together, the charged atoms captured on the capsule's disks of gold, sapphire, diamond and silicone are no bigger than a few grains of salt, but scientists say that's enough to reconstruct the chemical origin of the sun and its family of planets.

This clearly shows our need for gold in space as well as other "precious" metals and gems. This explains religion's description of the same in heaven/space. After all, religion is universally anti-wealth.

RELIGION IS HISTORY!!!
Missing Link Proof!

The alien skull shows the obvious mix between primitive man and himself producing us. We are the mystery! Religion began approximately 200,000 years ago when our big headed species started burying the dead and making artwork! This coincides with my theory like many others (Erich von Daniken and Alan Alford) that theorize intelligent beings from the sky manipulated primitive man and created this missing link in skull growth. The evidence shows that these religious gods are bigger headed than us and look like the stereotypical Roswell gray alien! This mix of their large head and primitive man's smaller one caused ours to jump the normal growth of evolution. Primitive man remained so for many millions of years without any evolution in knowledge or much change in skull size. The only logical explanation for the missing link and inexplicable sudden attainment of knowledge is religion itself. Not slow evolution, but a sudden scientific creation. The fossil evidence supports this reality! Religion is from Heaven or the sky! It is universal in the creation stories; it is clear that we are created to worship them or work for them. It is universal. This work or "purpose" can be traced directly to our first order of business. It was and is gold. It is universal. Gold is important for space travel. We are creating robots to assist in our space travels, which is only made possible with gold! This is universal! Even the robot itself is mostly composed of gold. It is the most resistant protector to the extreme conditions of space. The gold halo is the universal god symbol which is above the head. All gods are depicted as being able to fly. I am presenting the following religious symbols and their matching scientific counterparts as evidence to prove the creation of man was and is a scientific one. I propose that our true purpose is religion's pre-destined will; their pre-determined plan for us. Our true purpose is why they stay away on purpose. Please enjoy the exciting conclusion to my story. See for yourself how science is revealing an unfolding pre-determined plan for mankind that mirrors religion itself. Science today matches religion replacing the spirit-magic god with the alien!

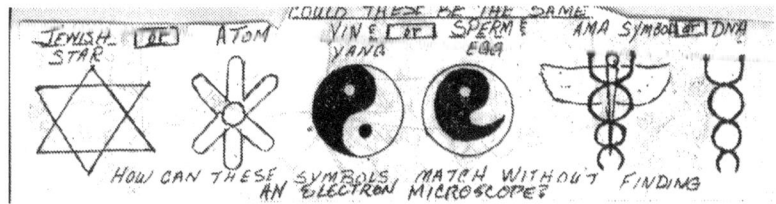

Symbols:
Matching Themes:

Religion	Science
Omnipresence	Space Program
God is Infinite	Atom is Infinite
Creation/Man	Robots
Saving Life	Improve/Saving Man's Life
Multiply	Ensure Propagation of Species
Manipulating Age	Genetics for Manipulating Age
One Mind	Evidence Rules
Levitation	Anti-Gravity
Mummification	Cryogenics
Mental Telepathy	Cybernetics
Spirit	Holograms
Disappearing	Invisibility/Teleportation
Mystery™	Biorhythm Feedback

I'm sure there are many more. So please read on as I "have" to get to work! I'm sure you'll get the point. I propose that religion is not only the best evidence that space is already conquered, but tells us plainly the answer to Fermi's paradox: " Why don't they contact us?"! The world just hasn't come together to scientifically answer it. But of course we all know the world is just now global and capable of destroying itself. These are the two pre-requisites for the end to be ushered in. Coincidence? Our true purpose explains their silence and the reason they stay away on purpose. It is for the good of science. However, this is the end of their tortuous silence. It is the only way to save our planet. It is the next step in their plan. The two witnesses torture the world with their prophecy! Our species is the most unnatural self-destructive species in the universe. Get ready for contact. The "Regeneration of Man", Second Coming, Mayan Golden Age.

18A Friday, December 20, 2002   THE TENNESSEAN   www.tennessean.com

DNA similarities make world seem smaller

Survey says any two people 99.9 percent identical

By LEE BOWMAN
Scripps Howard News Service

Although everyone's genetic makeup is unique, scientists have found that populations from different parts of the world still share more genetic similarities than had been thought.

The results of a computer analysis of DNA from individuals representing 52 populations around the globe, published today in the journal Science, make up the largest such global survey of genetic diversity, and should help studies of ancient human migrations.

Those surveyed were broken into five regions: Africa, Eurasia, East Asia, Oceania and the Americas. Differences among individuals within those groups accounted for 93%–95% of genetic variety, according to the international team led by Marcus Feldman, a professor of humanities and sciences at Stanford University.

Compare the genetics of any two people, and the matchup will be about 99.9% identical. The research team accurately pinpointed the ancestral content of virtually every individual from Africa, East Asia, Oceania and the Americas. ■

1. How could Hopi medicine man know of five races, let alone the alien god? See illustration on the next page.

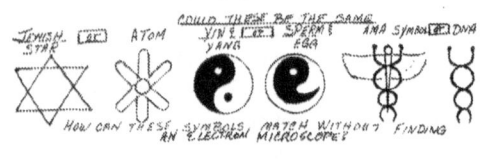

Alien statues found along banks of Jordan River 10,000 years old

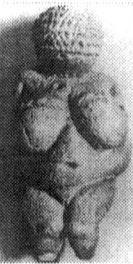

2. Notice alien head on mother goddess statue. This supports what gods look like in Genesis 6:4. Statue dated circa 5000 years old found in Jerusalem, Israel.

3. Notice asexual organs on alien statue. This confirms Yeshua's description of angels and explains why they don't give their hand in marriage. They must be androgynous.

Mother goddess statues represent the inevitable separation that was to occur. According to religion mankind is predestined. It happened because the "sons of gods" thought the daughters of men were pretty. Their "giant" offspring became men of great renown and all wickedness spread all over the earth. This exemplifies their lust for power due to their obvious oneness in looks and small size. And they must have considered themselves ugly. It took place during the mysterious time frame, of the last ice age approximately 13,000 years ago up to the beginning of the Jewish calendar, 4000 BC (6,000 years ago). These are found all over the earth. The alien headed one is from Israel circa 8000 years old. The round headed one is 30,000 years. It is called the Venus of Willendorf. The asexual alien statue was found along the banks of the Jordan River. Notice the circles as if they knew about chromosomes and DNA. These as all scientists agree were religiously important and found in every household. I theorize that like the pyramids and Easter Island giants they were left to stand the test of time to tell us what their gods look like and where they are, Aliens and space.

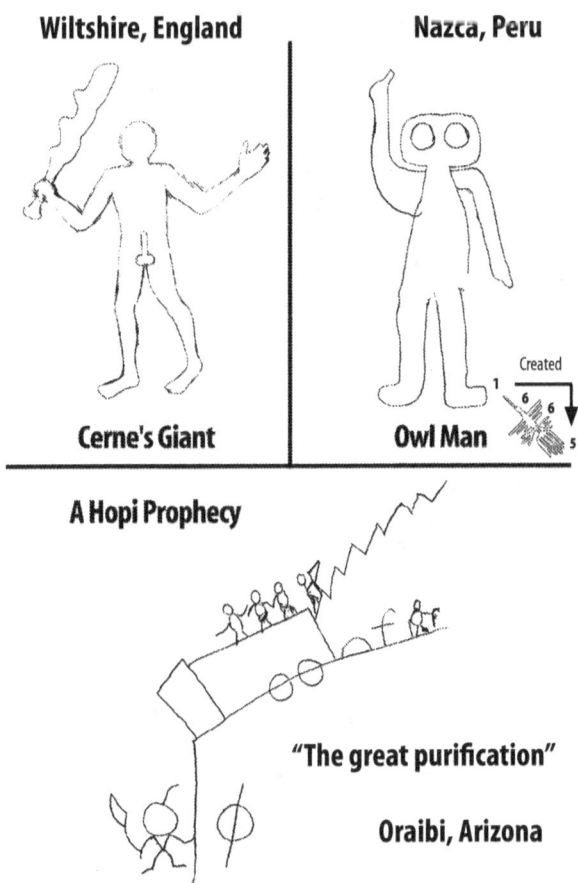

The Cerne's Giant tells us why they couldn't cohabit with their creation of modern man (we kill for sex) and when they will return. (Three humps on club indicate an impending nuclear disaster. Atom has three parts.) The Owl Man tells us where their gods live and what they look like, space and aliens. It has a hummingbird which represents fertility pointing at it. Count the appendages and see connection to biblical number of man 666 and five races. The Hopi prophecy again shows us an alien god (big headed guy) saving the earth from destruction and recycling the majority of man up, obviously to another planet. Again how did the shaman know of the five races of man let alone an alien God? Notice similarity of box carrier to today's truck trailer. Feather on head indicates gods' ability to fly. See saucer attached to his arm.

The Five Faces of Man

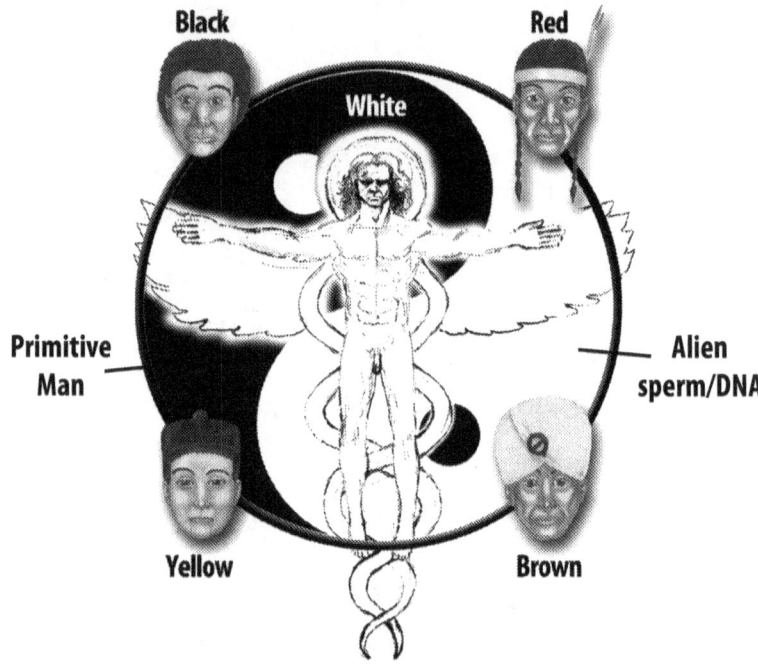

The different colors and facial structures indicate the competition to make the "prettiest" human. This "prettiest" factor is evident in Genesis 6:4 and the fall of the angel story!

Looks give us power over one another. All religions have a fall from heaven and earth being one, void and without form, to being separated. The gods/angels live in space—Earth becomes prison. This resulted from a power struggle. The biblical account gives us two creations. Nature created primitive man and then the gods/angels/aliens scientifically created modern man as a worker. Modern man is the mystery. The evidence universally points to mining gold. This started 100,000 years ago and continues to this day! The matching yin and yang and AMA symbol to the science symbols of the sperm/egg and DNA reflects our scientific creation. Even the biblical account describes a scientific process both for the man and the woman. The woman's creation is from man and he is anesthetized. Ultimately, I theorize two ongoing infinite creations: Nature's gods/angels/white sperm/DNA and us from primitive man/black sperm/DNA. We are the mystery!

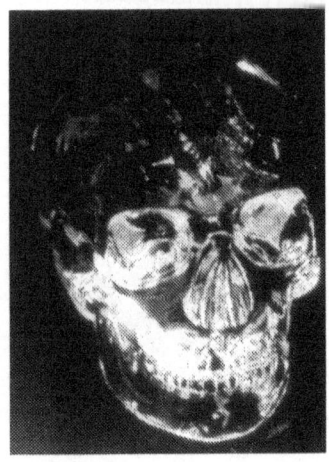

One of thirteen Mayan quartz crystal skulls. They are anatomically perfect, ancient and show no tool marks. Legend has it that they hold information that will solve our mystery which is where are the people/gods who made us and them? This crystal today is used for its electrical conduciveness "piezo electricity" and storage of information on computer chips.

This is where I propose they are. This is a statue from Easter Island that is looking up and appears to have a flying saucer on top of his head. They are in these religious fiery chariots of the sky.

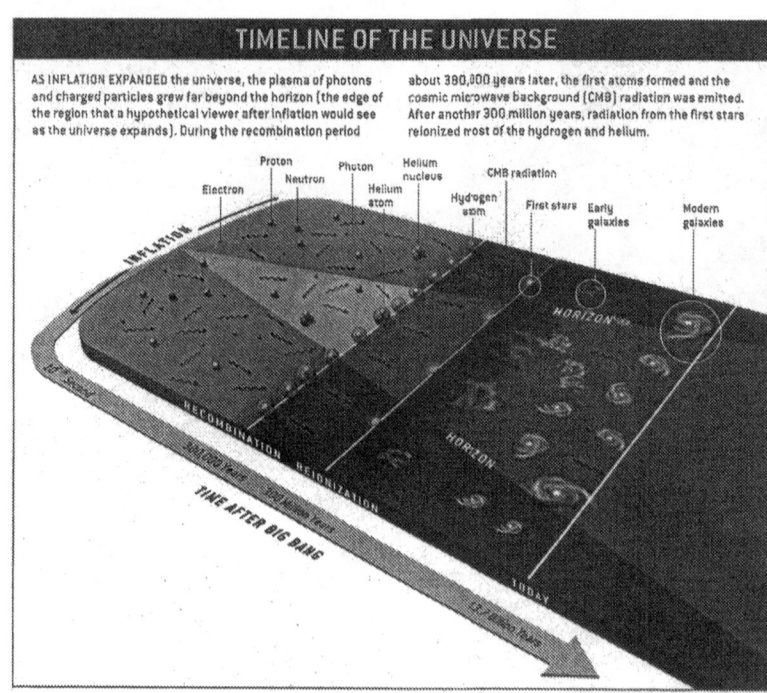

This is a perfect example of macrocosm science. The smallest parts mirror the whole. Atoms, photons, solar systems galaxies all resemble the universe itself. It is mostly space. The ancient geometry supports this theory/reality. I was amazed at how much a woman's egg looks like the sun (magnified) and when the sperm penetrates it the outer shell grows a green growth that becomes the placenta. The earth would only grow green vegetation from photon penetration. Photons look like sperm. Atoms look like suns, these look like eggs! Anyway, The problem with a beginning to our universe is that it is infinite. Only matter has a "beginning and ending". But this is an oxy-moron because atoms make matter and though one form ends it doesn't stop existing, another one just begins. This is all about image! Mind over matter and to be free of matter constraints, we must free ourselves from the matter. For space travel/freedom it is literally what we have to do. Free ourselves from the inevitable invisible eater of matter. GRAVITY! Matter itself. Does all this really matter? To be free it does. Flying is the ultimate freedom!

Finally my ex-religion's fraud!

Chapter 24   **Chapter 25**

Handwritten note (exhibit a):

HERE is the proof of fraud. The Jehovah witness actually worship Jehovah. The gang but don't call themselves Yahoeh witnesses. Most important is they don't recognize that Jeshua is the correct translation from Yeshua. It is in the book of EZRA 3rd chapter. It is from Yeshua. Why isn't this same in the N.T.? It is easy to prove that Jesus, Jehovah, James, and Christ are all fraudulent, added words. (they don't all have new words) We even have the correct names in the bible. That is if you accept the changing of J's to J. In this code with the Jehovah's witnesses, they actually choose Jehovah over Yahweh! The four correct translations that are in the bible are Jeshua, Yahweh, Jacob, a messiah. Please hold them accountable for accuracy. This is not interpretation.

Sincerely,
— Mike Brumfield

1. "Where is God?" is the $64,000 question. I thought God is omnipresent! That means everywhere like the atom! They say he's all alone in space!

2. They say he's not lonesome but he's all alone. Then they say he creates for others. That's loneliness.

3. Finally they say he creates a heavenly organization of "spirit" sons like "himself". Why not daughters? And now they skip the fall of the angels story.

4. Last but not least. This is the proof that "religion", at least the Jews, make these angels and god "spirit" not flesh and blood. This is the ultimate cover-up. What if they come back and are the aliens? WWYD?

5. Proof that the J.W.'s discredit science! And ironically the scripture above James 2:9 makes their god a hypocrite. He has a favorite, yet forbids it. The chosen race of the Jews. No wonder people revere the Jews!

6. Finally they say the "Devil" is working through the U.N.! Don't they want a United Earth?

Modern-Day "Encounters" With Angels and Aliens

Many people today claim that they have seen angels and spoken with them. Others say that they have had contact with aliens from other worlds. The book *Angels—An Endangered Species* lists the similarities between these accounts, claiming that both may have a common explanation.* Following is a summary of some similarities listed in the book.

1. Both angels and aliens come from other worlds.

2. Both are advanced life-forms, either spiritually or technologically.

3. The friendly variety are youthful and beautiful in appearance, and they are kind and full of compassion.

4. Both have little trouble with language, speaking clearly in the language of the listener.

5. Both are masters of flight.

6. Appearances of both angels and aliens are accompanied by brilliant light.

7. Both appear fully dressed, commonly in either robes or close-fitting tunics. White or blue are favorite colors.

8. Both are usually the same height as humans.

9. Both express concern about the plight of humanity and the planet.

10. The evidence of both alien and angelic encounters is the testimony of the beholder.

The explanation common to both is that wicked spirits, or demons, are evidently behind many such "encounters." As the Bible says, "Satan himself keeps transforming himself into an angel of light." (2 Corinthians 11:14)—See Awake!, *July 8, 1996, page 26.*

★ **ANSWER**

Erich von Däniken

...to count the pages of his metal library, but I accept his estimate that there might be two or three thousand.

The characters on the metal plaques are unknown, but if only the appropriate scholars were told of the existence of this unique find now I am sure that they could be deciphered comparatively quickly in view of the wealth of possibilities for comparison.

No matter who the creator of this library was, nor when he lived, this great unknown was not only master of a technique for the "mass-production" of metal folios in vast numbers—the proof there—he also had written characters with wh... he wanted to convey important information beings in a distant future. This metal library w... created to outlast the ages, to remain legible fo... eternity.

Time will show whether our own age is seriously interested in discovering such fantastic, awe-inspiring secrets.

Is it prepared to decipher an age-old work even if it means bringing to light truths that might turn our neat but dubious world picture completely upside down?

Do not the high priests of all religions ultimately abhor revelations about prehistory that might replace belief in the creation by knowledge of the Creation?

Is man really prepared to admit that the history of his origin was entirely different from the one which is instilled into him in the form of a pious fairy story? CALLED RELGT'S SPIRIT WORLD!

This is why we must prove religion's origin. The Jehovah Witnesses are not helping to make peaceful contact! They don't even realize that their answer proves mankind is religion's devil and reincarnation! (Keeps transforming).

A ridiculous depiction of God by Jehovah's Witnesses. Sadly, this is universal. Religion is anti-wealth, making this even more insulting.

Christ is no longer a babe but the powerful King of God's Kingdom

How to identify true religion
What good fruit should true religion produce?—Matthew 7:17.

"Something Cannot Come From Nothing"

■ **KENNETH LLOYD TANAKA** PROFILE: I am a geologist presently employed by the U.S. Geological Survey in Flagstaff, Arizona. For almost three decades, I have participated in scientific research in various fields of geology, including planetary geology. Dozens of my research articles and geologic maps of Mars have been published in accredited scientific journals. As one of Jehovah's Witnesses, I spend about 70 hours every month promoting Bible reading.

Using science to prove God and make themselves look ridiculous with their quote, which disproves their God. (read about this in my story)

Could I be this Michael and the war about our evil species? Notice Tibetan religion's heavenly war also involves one third being rebellions! Their Sixth Stanza reads like modern science. The final admonition is to learn the "correct" age of the "small wheel." This is the atom and us. We are atoms/adams "appearing and reappearing continuously."

Atoms are infinite!

> ### The Gold of the Gods
>
> stars of God: I will sit also upon the mount of the congregation, in the sides of the north."
>
> But we also find an unmistakable reference to strife in heaven in the New Testament. Revelation xii, 7-8, reads:
>
> "And there was war in heaven: Michael and his angels fought against the dragon; and the dragon fought and his angels,
>
> "And prevailed not; neither was their place found any more in heaven."
>
> Many of the ancient documents of mankind mention wars and battles in heaven. The Book of Dzyan, a secret doctrine, was preserved for millennia in Tibetan crypts. The original text, of which nothing is known, not even whether it still exists, was copied from generation to generation and added to by initiates. Parts of the Book of Dzyan that have been preserved circulate around the world in thousands of Sanskrit translations, and experts claim that this book contains the evolution of mankind over millions of years. The Sixth Stanza of the Book of Dzyan runs as follows: *[handwritten: Like sixth day beginning in Bible]*
>
> "At the fourth (round), the sons are told to create their images, one third refuses. Two obey. The curse is pronounced . . . The older wheels rotated downward and upward. The mother's spawn filled the whole. *There were battles fought between the creators and the destroyers, and battles fought for space;* the seed appearing and reappearing continuously. Make thy calculations, o disciple, if thou wouldst learn the correct age of thy small wheel."

Satyr-comedy-Aristophanes: Lysistrata

[handwritten: Greek "scientist" (Philosophy crossed out)]

Socrates: 469-399 BCE
  Socratic method/dialectic method
  "The unexamined life is not worth living."
Plato: 429-347 BCE, the Academy *(Aristotle)*
  "Until philosophers are kings or the kings and princes of the world have the spirit and power of philosophy...cities will never cease from ill, nor the human race."
Aristotle: 384-322 BCE, the Lyceum -peripatetic
  "Plato is dear, but truth is dearer."

[handwritten:] Albert Einstein "I wonder if nature did not always play the same game"

Erich Von Daniken "I theorize that Alien intelligences must have been the same as homo sapiens or very much like him!"

Mike Brumfield: "I theorize the Roswell alien is the scientific creator of modern man/homo sapiens everywhere in the universe for the "soul" purpose of power and working for/worshipping him!

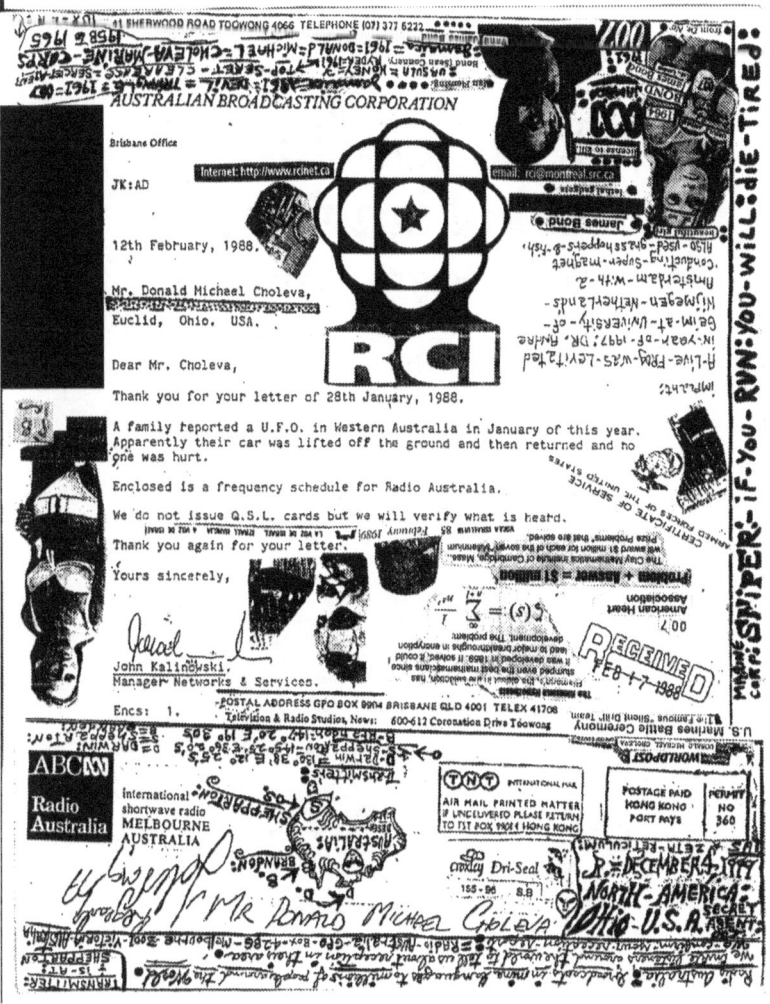

Blackened box in upper left corner had Marine Corps sniper stamp. This is a death threat of which I contacted the police. It came in the mail!

Clearly, scientists want cloak of invisibility

If they could only see a way to make it ...

By ANDREW BRIDGES
Associated Press

WASHINGTON — The key to creating a Harry Potter-like invisibility cloak lies in manmade materials unlike any in the Hogwarts School of Witchcraft and Wizardry, researchers say.

They're laying out a blueprint for turning science fiction into reality. And they say that, in theory, nothing's stopping them from making such a cloak.

Well, almost nothing: They still need to perfect the manufacture of those exotic materials with an ability to steer light and other forms of electromagnetic radiation around a cloaked object.

"Is it science fiction? Well, it's theory and that already is not science fiction. It's theoretically possible to do all these Harry Potter things, but what's standing in the way is our engineering capabilities," said John Pendry, a physicist at the Imperial College London. Details of a study that Pendry co-wrote are in Thursday's online edition of the journal Science.

"This is very interesting science and a very interesting idea, and it is supported on a great mathematical and physical basis," said Nader Engheta, a professor of electrical and systems engineering at the University of Pennsylvania who has done his own work on invisibility using novel materials called metamaterials.

Pendry and his co-authors also propose using metamaterials because they can be tuned to bend electromagnetic radiation — radio waves and visible light, for example — in any direction.

A cloak made of those materials would neither reflect light nor cast a shadow.

Instead, like a river streaming around a smooth boulder, light and all other forms of electromagnetic radiation would simply flow around it. An onlooker would appear to peer right through the cloak, with everything inside it concealed.

Early versions that could mask microwaves and other forms of electromagnetic radiation could be as close as 18 months away, Pendry said. He said the study was "an invitation to come and play with these new ideas."

"We will have a cloak after not too long," he said. ∎

Invisibility is the Holy Grail of Science. Also, it is at the core of religion and the modern-day UFO phenomenon.

MODERN PHYSICS

THE BASIC ELEMENTS OF MATTER

What is **matter**?

Matter is anything that takes up space and has mass (or weight, which is the influence of gravity on mass). It is distinguished from energy, which causes objects to move or change, but which has no volume or mass of its own. Matter and energy interact, and under certain circumstances behave similarly, but for the most part remain separate phenomena. They are, however, inter-convertible according to Einstein's equation $E = mc^2$, where E is the amount of energy that is equivalent to an amount of mass m, and c is a constant, the speed of light in a vacuum.

In 1804, the English scientist John Dalton formulated the atomic theory, which set out some fundamental characteristics of matter, and which is still used today. According to this theory, matter is composed of extremely small particles called atoms, which can be neither created nor destroyed. Atoms can, however, attach themselves (bond) to each other in various arrangements to form molecules. A material composed entirely of atoms of one type is an element, and different elements are made of different atoms. A material composed entirely of molecules of one type is a compound, and different compounds are made of different molecules. Pure elements and pure compounds are often referred to collectively as pure substances, as opposed to a mixture in which atoms or molecules of more than one type are jumbled together in no particular arrangement.

341

People ... please learn that matter is energy and infinite! This explains the $E = mc^2$ of energy. Atoms can be neither created nor destroyed!

Question: Is this Yeshua's/Jesus's sign for contact... nuclear war?

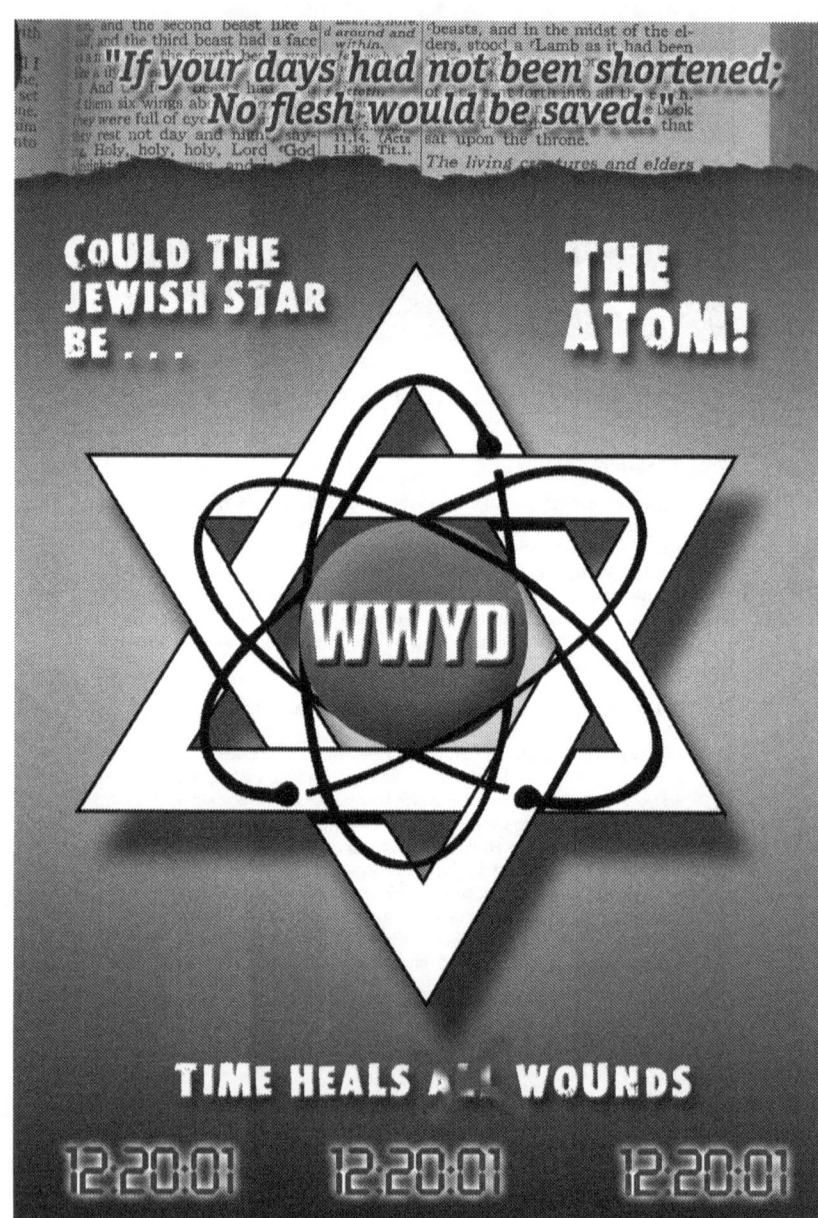

Proof

DVD 19.95-BOOK 12.95
SEND CHECK OR M.O.
MIKE BRUMFIELD
1066 GOLDEN HERREN RD
SPARTA, TN 38583
ORDER ONLINE AT
UFOSOVERPHOENIX.COM
931-261-6697

2005

History Researcher Mike Brumfield examines Ancient Evidence/ Art/ Religion giving his opinion of why **Extraterrestrials** won't make open contact with the human race. Featured In New Book "2012 Gold's History solves Mankind's Mystery" Also Featuring Jeff Willes Famous Video Footage Recently Previewed On The Travel Channel. Matches Johnny Cash & Easter Island !

Crop Circle

Yin and Yang or Atomic Propulsion

Jeff's

FACT
Ancient Religious Statue- "Giant Heads" of Easter Island

OPINION
It is looking up & telling us where their GOD is: UP/ HEAVEN/ SPACE
What they're in: Fiery Chariot/ Flying Saucer. They Look UP because all religions await "SECOND COMING from the SKY"
HEAVEN IS "UP" / SKY / SPACE!!

proof

WHY THEY DONT SHOW UP

MIKE BRUMFIELD

Easter Island Solved

Proof

Introducing Jeff and Mike they find it!! Jeff gives you current flying saucer footage that matches Johnny Cash sighting in 1980 & Easter Island *(back cover)*

Mike Brumfield REVEALS
Scientific Religous Evidence
"Why They Don't Show And When They Will"

More inside...

Jeff's Johnny's
Matching Saucer's!

Real photo of Johnny Cash in 1980

Solve's Easter Island Mystery And More!

Famous Painting 1400 A.D. of a Flying Saucer

Flying Saucer Matches Also!

Man looking "UP" at it..

Close up View

The Jeff and Mike Show—Flying Saucer Hunters

WHY THE BLANK DON'T THEY CARE

VIDEO 200

We were also shocked to look "UP" and see them. Did these ancient religious "chariots of fire" create religion's stories, "heavenly beings in the sky" and their universal halo symbol "UP" above their head? It is a saucer shape!

HEADS OF EASTER ISLAND LOOKING "UP"! THIS EVIDENCE MATCHES
Ancient Flying Saucer Statue

Letter to *UFO Magazine*, February 20, 2006, about Jeff's proposal to find flying saucers. They didn't bite, and we filmed them during trip. The still didn't bite even after contacting them with our footage to prove we found them!

Bill:

I have enclosed a video for you to review. I recently talked with you about the Johnny Cash photo and emphasized how this could garner serious attention for a huge breakthrough discovery. The flyer clearly shows a match between Jeff's flying saucer and the alleged hat of Easter Island. This also matches Johnny's saucer which is on the cover of the video. Please bear with the amateur quality of my video. My narration/explanation of why they don't show is one hour long. Then the 2003 video footage of Jeff's flying saucers begins. I also included the 1952 White House incident. This alone should demand attention. I will be coming to LA and approach the *Times* with my breakthrough discovery in man's quest for contact. The head of Easter Island is clearly looking up and has a flying saucer on top of its head. This answers the questions that Barbara Walter's just posed to all esteemed religious scholars of the Earth including the Dalai Lama, "Where is heaven?" My breakthrough religious scientific discovery of the world's largest ancient statue of a flying saucer confirms all ancient writings as well as today's space pursuit. Heaven is up, space is the final frontier. It is already conquered like religion confirms. The owl man of Peru shows us the same message. He is pointing up! The owlman and Easter Island statues are bald headed big-eyed statues like the Roswell alien. There are many depictions in ancient artworks of aliens and flying saucers that confirms this reality. I am sending also a cover of my latest book showing some of these that can't be refuted. I hope you will consider contacting me for an interview. We can't believe that nobody is taking this seriously. That is why we titled our next video "Why the Blank Don't They Care." I make it clear in the video why they don't show. The evidence speaks for itself. We are mining gold for them, since our beginning. Gold is important for space travel. Our creation is scientific and can't be stopped. There is no spirit magic and people get addicted to beauty of the flesh which gives us power over one another. The Easter Island statues all look the same. It would be great to have your magazine chronicle our journey on getting some attention for this discovery. We will prove that flying saucers are here now by videoing them together. This will happen after I leave LA on my return trip to Nashville. My partner Jeff lives in Phoenix and films them regularly. He promises me that we will see one. I believe him. Do you? Funny thing, that religion's universal contact story is about believing. We propose that this flying saucer evidence will answer that question. And yet, the story goes that most won't. We are entrenched in a spirit world belief system. Could religion be a direct result of primitive man's contact with flying saucers and aliens. They have both universal traditions, a gold halo symbol and bald heads.

Sincerely,
Mike

People, crimes and mysteries are solved today through the scientific method of matching evidence. The perpetrators of religion live in the sky. We are now videoing flying saucers everywhere and we now live in the sky! Could flying saucers and a real flesh and blood alien species have created religion as the matching evidence suggest? If you can open your mind to accepting evidence it is easy to "SEE" how advanced technologies would have created primitive man's spirit magic traditions. However, fortunately enough the foundation of his story though has never changed. These people live in the sky. The evidence does show "HEAVEN IS SPACE, UP!"

AND THIS IS THEIR SYMBOL . . . .

The gold halo above their head is their symbol! It looks like a flying saucer! What they look like is the ultimate evidence that will solve our mystery and Fermi's Paradox. The statues on the next page clearly answer the question of who and what primitive man's god looked like. Why they created our species is obvious. It also matches their story. They desired beauty of the flesh. This gives them "greatness" over one another. Coincidence? "Religion cover-up" is the title to my first book.

This is the "head/person" that should be under the gold halo. Funny how the book cover shows the shroud which looks like space beneath it. Because it exemplifies what the cover-up in religion is all about. What heaven's occupants look like? And where are they? The evidence says clearly that they are aliens UP IN SPACE! If you don't follow evidence or can't imagine a new discovery of an ancient relic that could solve our mystery, then please watch "The Planet of the Apes." The apes ignore scientific evidence over their religious stories of God creating them, just like my story today! Why the blank don't they care! Check out my video. Are we all addicted to outward beauty? Hell, yeah!!! Can't, couldn't do anything. I've always said instant creation/magic wasn't possible, but if it is, I guarantee it to be scientific. And then, it couldn't make this universe perfect. Babies are being raped. Finally, people . . . please answer for yourself why anybody would let this happen if they could stop it. I promise you, there's only one logical answer . . . they can't.

Please ask yourself one question . . .
Do you like beauty of the flesh?

Oh, and by the way,
if you thought
What The Bleep Do We Know
was good, wait until
you see my next film . . .

What the Blank Is Their Problem?

This last picture will
"drive my point" . . .
home!

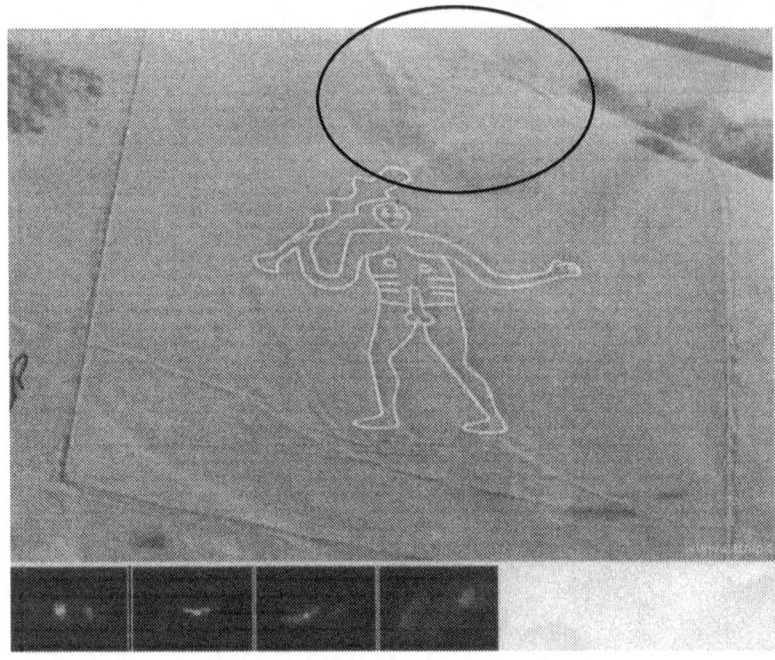

This is the Cernes Giant of Cerne, England. Legend has it that it is a Saxon god. My conclusion, based on the evidence I've presented, is that the god of primitive man is an alien and his fiery chariot is a flying saucer. The images below the giant are from video that I shot in Phoenix, Arizona, March 6, 2006. It is a flying saucer. It also matches the mound circled/squared above the head of this Cernes Giant/god. The most compelling matching evidence is the shape of the mound, a ring within a ring, and the bald head and big eyes of the giant. The craft matches my craft as can be seen at the following exhibit's website, and the head of this giant matches the aborigine alien-looking god, as well as all others. My last word on this evidence issue, to religious people and the rest of the world, is the bald-headed traditions of all religions and looking UP! This relief carving is another example of part alien/part man with a fiery chariot "UP" above his head!

Get the point!

rense.com

UFO Performs Acrobatics Over Phoenix

3-15-6

3 witnesses observed the videotaping of a UFO performing aerial stunts so astonishing it made it on the Channel 3 News in Phoenix. Once again, Jeff Willes' constant skywatching pays off big time...

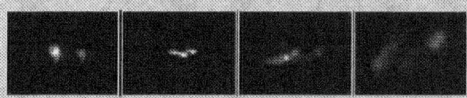

To view part of the video click here
(Windows Media Player file - wmv)

Report from Jeff Willes of NFO

"I was out in the back yard skywatching late on 3-6-06. At 12:00 am one of the others skywatching with me saw a UFO come up from the trees about a mile away and then go down again. We ran up to the top of the roof leaving our tripods on the ground, forgetting them in all the action. When we got on the roof we saw not one but two UFOs shooting way up in the sky and then coming back down behind the trees. The UFOs shot up and came back down 3 times. One of the craft would fly over upside down and then dive down. It did this 3 or 4 times. I took the tape to KYVK Channel 3TV here in Phoenix. They aired the footage on 3-10-06. They showed how the object flips over. I have been videotaping UFOs sense 1995 and have never seen anything like it."

Jeff Willes
www.ufosoverphoenix.com

I was the other witness along with Jeff and his wife. The daytime footage I shot is on the back cover of this book. I spent three days and videotaped well over ten flying saucers. Even the news reporter commented that the behavior of our saucer footage definitely proved it was not any kind of aircraft known to man. You can view the footage at **ufosoverphoenix.com**!

I dedicate the ending to Jocelyn Elders. She advocated teaching acceptable masturbation in schools. The Cernes Giant *drives my point home*. I am sexually addicted to power/beauty of the flesh. I think we all are. If you are not open-minded to an infinite universe, and an ugly intelligent species like the aliens, which all look the same, creating our species for power through unique beauty of the flesh, then I don't know what planet you live on. Because I live on a planet that is ruled by beauty of the flesh, *AND WE HAVE ALIEN/FLYING SAUCER EVIDENCE!*

I would like to especially mention my great friend and editor, Jan. She has forced me to reconcile destiny versus chance. We have talked about karma many times, and I don't believe in it, because of freak tragedies. Babies are being raped right now, people. This is a fact. She now feels that the futuristic idea of creating our evil-natured power-hungry species is the only thing that makes karma logical. I will attempt to do this in my next book. Maybe we have lived before; maybe we will live again . . . who knows? Right now, all we know is the atom supports this reality as well as all the scientific evidence/religious stories. Please see the following recent article from The Associated Press on the next page. It speaks for itself.

(Religion is universally a cycle of birth, death, and this repeats itself infinitely. Matter goes through an infinite cycle of changing form.)

"Big Bang Theory disputed

Space and time go on forever and the so-called Big Bang said to have started the universe is actually part of a repeating cycle, according to a new paper that challenges conventional wisdom in physics. Infinite space and time would contradict the generally accepted notion of a universe expanding abruptly out of nothing 14 billion years ago, said Neil Turok, a professor of mathematical physics at the University of Cambridge in England.

Turok wrote the paper in the journal *Science* with Paul Steinhardt of Princeton University. An ongoing expansion and contraction is more likely than a Big Bang as supported by Cambridge physicist Stephen Hawking and other cosmologists building on the work of the late Albert Einstein, Turok said in a telephone interview."

In the words of Michio Kaku, "Let this investigation begin."

The future will reveal the technical reality of redefining religion's spirit/soul story and reincarnation. The ability to create humans for the purpose of power lies ahead. It is all about beauty of the flesh! If you can't envision science replicating our memories and transplanting/downloading them into newly created bodies, then you aren't following today's science. We are on the threshold of mastering memory cell transplantation; not to mention, brain augmentation and cybernetics. Religion is a universal story of "smarter" people living "UP" in the sky, watching "US." Funny, huh? Human drama is also our number one form of entertainment as well!

Please buy yourself a copy of my next book,

THE DISCOVERY

www.ingramcontent.com/pod-product-compliance
Lightning Source LLC
Chambersburg PA
CBHW031245290426
44109CB00012B/446